T0205489

Sustainable Textiles: Production, Processing, Manufacturing & Chemistry

Series Editor

Subramanian Senthilkannan Muthu, Head of Sustainability, SgT and API, Kowloon, Hong Kong

More information about this series at http://www.springer.com/series/16490

Subramanian Senthilkannan Muthu
Editor

Microplastic Pollution

Editor
Subramanian Senthilkannan Muthu
Sustainability
SgT Group and API
Kowloon, Hong Kong

ISSN 2662-7108 ISSN 2662-7116 (electronic)
Sustainable Textiles: Production, Processing, Manufacturing & Chemistry
ISBN 978-981-16-0299-3 ISBN 978-981-16-0297-9 (eBook)
https://doi.org/10.1007/978-981-16-0297-9

This Springer imprint is published by the registered company Springer Nature Singapore Pte Ltd.
The registered company address is: 152 Beach Road, #21-01/04 Gateway East, Singapore 189721, Singapore

This book is dedicated to:

The lotus feet of my beloved Lord Pazhaniandavar

My beloved late Father

My beloved Mother

My beloved Wife Karpagam and Daughters—Anu and Karthika

My beloved Brother—Raghavan

Everyone working with Micro Plastics Pollution to make our planet Earth SUSTAINABLE

Contents

About the Editor

Dr. Subramanian Senthilkannan Muthu currently works for SgT Group as Head of Sustainability, and is based out of Hong Kong. He earned his Ph.D. from The Hong Kong Polytechnic University, and is a renowned expert in the areas of Environmental Sustainability in Textiles & Clothing Supply Chain, Product Life Cycle Assessment (LCA), Ecological Footprint and Product Carbon Footprint Assessment (PCF) in various industrial sectors. He has five years of industrial experience in textile manufacturing, research and development and textile testing and over a decade's of experience in life cycle assessment (LCA), carbon and ecological footprints assessment of various consumer products. He has published more than 100 research publications, written numerous book chapters and authored/edited over 95 books in the areas of Carbon Footprint, Recycling, Environmental Assessment and Environmental Sustainability.

Effect of Textile Parameters on Microfiber Shedding Properties of Textiles

S. Raja Balasaraswathi and R. Rathinamoorthy

Abstract Fast fashion is one of the recent changes in the apparel and fashion industry. The fast-fashion process enables mass customers to adapt to the current trend in a short period. To reach the customers at the earliest time at an affordable price, the manufacturing industries use cheaper raw materials in various processes of manufacturing. Synthetic textiles like polyester, polypropylene, and acrylic are the common fibers used in the fast-fashion items. These fibers are identified as a major source of microfiber pollution. The synthetic fabric sheds microfibers during wearing, laundry, and disposal. A recent study estimated that the synthetic clothing sale would reach 160 million tons in 2050 and subsequently 22 million tons of microfiber added into the ocean between 2015 and 2050. The microfiber shedding of textiles is mainly influenced by the different production and fiber parameters. The chapter aims to elaborate on the role of different synthetic fibers and their shedding properties. Similarly, the influence of fabric manufacturing methods like weaving and knitting also detailed for a better understanding of shedding. Further, this chapter details the fabric parameters like fabric thickness, fabric density, structure type, and their influence on microfiber shedding. The last part of the chapter deals with the influence of different finishing methods to control or reduce the microfiber shedding from synthetic fabric.

Keywords Microfiber shedding · Fiber type · Yarn type · Twist and hairiness · Fabric types · Abrasion resistance · Pilling · Tensile strength · Finishing process

1 Introduction

In the past few decades, plastics have become an essential, unavoidable need in all aspects of day-to-day life. Plastics hold a Global Market size of 568.9 billion USD in the year 2019 and the estimated compound annual growth is at a rate of 3.2% during 2020–2027 [1]. Plastics that are less than 5 mm in length fall under the

---

S. Raja Balasaraswathi · R. Rathinamoorthy (✉)
Department of Fashion Technology, PSG College of Technology, Coimbatore, India
e-mail: r.rathinamoorthy@gmail.com

S. S. Muthu (ed.), *Microplastic Pollution*, Sustainable Textiles: Production, Processing, Manufacturing & Chemistry, https://doi.org/10.1007/978-981-16-0297-9_1

microplastic categories which are very difficult to be traced [2]. These can be in the form of fragments, granules, or fibers [3]. Based on these sources, microplastics are classified as primary microplastics and secondary microplastics [4]. The first one is the manufacturing and usage of plastics of smaller sizes (Primary microplastics) and another is due to the fragmentation or degradation of larger plastics into smaller ones (Secondary microplastics) [2]. These microplastics should also be addressed and it seeks even more attention than macroplastics since traceability, prevention, and removal of microplastics are extremely challenging. The microplastics are generated from various sources including erosion of tyres, synthetic textiles, marine coating, road markings, personal care products like cosmetics and plastic pellets [4].

Among various sources, synthetic textiles play a dominant role in microplastic emission. Textile materials can be a potential source of microplastics in the form of short fibers. Synthetic fibers hold an irreplaceable place in the apparel and textile industries. The ultimate properties and performance of synthetic fibers get such a position in the textile field. The dominance of synthetic textiles in the industry keeps on increasing and it accounts for around 63% of total fiber production [5]. The recent accelerator of the global fashion market 'fast fashion' is one of the major reasons for the higher consumption of synthetic textiles. Fast fashion focuses on faster production at affordable prices. And synthetic textiles will be a good choice in achieving the low cost and quick production concept of fast fashion. Microplastic fibers which shed from synthetic textiles are considered as the main source of microplastic pollution. About 20–35% of microplastics found in the marine environment resemble the fibers which are similar to those used in the apparels [6]. In the study by Boucher and Friot to estimate the sources of microplastics in the marine environment, it has been reported that the synthetic textiles contribute more among various sources and it accounts for 35% of total sources identified [4]. Figure 1 represents various sources of microplastics from the literature. This confirms the important role played by

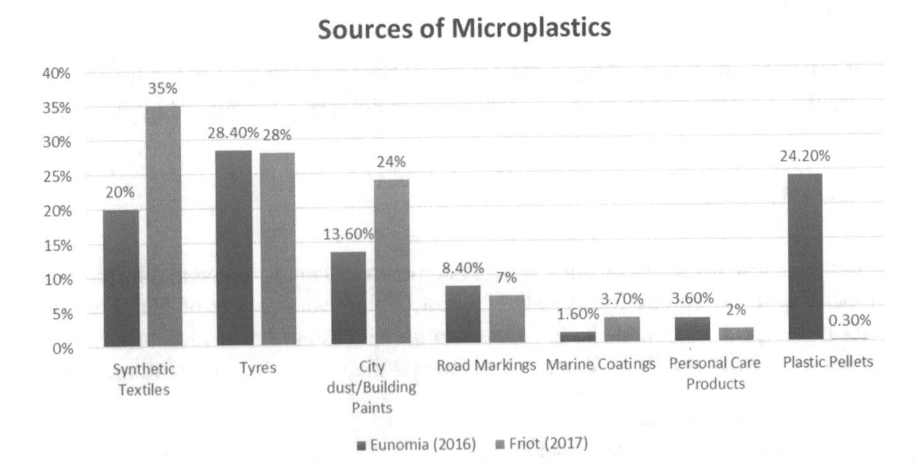

Fig. 1 Sources of microplastics in the marine environment [Authors own representation]

synthetic textiles in microplastic pollution. It is estimated that annually around 0.19 million tons of microfibers are entering the marine environment [7]. The increasing trend of synthetic textile production and consumption will proportionately increase the microfiber emission and it is estimated that the microfiber release into the marine environment will reach up to 22 million tons by the year 2050 [8].

Various researches have been made to understand the presence of microplastics in the environment and to analyze their characteristics to get a clear idea of their source. Microfibers are found to be present in different levels of the environment due to their varied densities [9]. Microfibers are found to be present in various samples that are taken from surface waters [10], sea ice [11], and even in the atmosphere [12]. Browne et al. collected 18 samples of sediments in the coastal region of six continents and they found microplastics in all the samples. The collected microplastics showed a close resemblance to the synthetic microfibers that are collected from the sewage treatment plant which confirms that the coastal regions are contaminated with microplastic fibers which are originated from the synthetic textiles [13]. The presence of microplastics in the sea ice in the arctic region has also been studied and the characterization of collected microplastics has shown that rayon accounts for a maximum of 54% which is then followed by polyester and nylon contributing 21% and 16%, respectively [11]. The other researcher has found microfibers in the river water samples. It has been estimated that around 300 million individual microfibers are being discharged along the surface of the river (Hudson River, USA) [10]. The research made in Saigon River also confirms the presence of microfibers in the river water. They noticed a huge proportion of synthetic fibers which contributes around 92% of total microfibers, where polyester being the dominant one with a 70% contribution [14]. The sewage effluents could be the potential source of microplastics in the freshwater systems and marine environment. A study on the effectiveness of wastewater treatment plants has concluded that sewage effluents have a variety of microplastics from various sources. The characterization of microplastics has shown that flakes hold a huge proportion of 67.3% which is followed by fibrous particles which account for 18.5%. They have also reported polyester as the most prominent polymer followed by polyamide with each contributing 28% and 20%, respectively, in the collected microplastics [15]. In addition to this, microfibers are also found in the air as the size was so smaller, and unfortunately, we are inhaling microfibers too [16]. A study on the analysis of microfibers in the air sampled in indoor and outdoor sites has confirmed the presence of microfibers in the atmosphere and a considerable number of microfibers ranging from 1.0–60.0 fibers per cubic meter are found. Another interesting fact to be noted here is the nature of fibers. Out of total microplastics found, around 33% of fibers found were synthetic fibers and they can add on to the microplastics [17].

Through various studies on microplastics' prevalence in the environment, it is confirmed that the synthetic textiles that are shedding microfibers are one of the major contributors to microplastic pollution. This arises the need for a detailed analysis of textiles throughout their lifetime to understand their potentiality of releasing short fibers which can add on to the microplastic load in the environment. This chapter tries to address the existing gap in the research area by providing the role of textile characteristics on synthetic textile microfiber shedding or releasing ability in different situations. Especially, the major focus is provided on the laundry process as most of the release occurs during consumer washing. This becomes essential as most of the recent researchers are alarmed about the microfiber contamination of the aquatic system.

2 Microfiber Shedding Mechanism

The disengagement of loose or damaged or broken fibers from the surface of textile materials is often referred to as the microfiber shedding of textile materials. Generally, when the textile materials are subjected to mechanical stress, the abrasion on the surface of the materials leads to fiber damage or breakage [18].

Even the handling of materials while using can also lead to fiber breakage as the fibers can be broken down by the application of repeated small or moderate loading rather than a single excessive force. Tensile fatigue and flex fatigue can lead to fiber failure. Not only breakage, but fiber failure can also be in the form of splitting and peeling. The reason for fiber breakage or damage is not limited to mechanical actions but the fibers can also get degraded due to chemical actions [19]. This can often happen during wet processing where fabrics are subjected to a wide variety of chemicals [20] and also during domestic laundering where detergents [21] and other laundry additives are used. This fiber damage or breakage leads to fiber shedding as these short fibers can get disengaged from the surface of the textile materials. Moreover, during the yarn production process, the fibers will be subjected to various mechanical stresses. As a result of this, fibers can be cut or broken down and the small fragments of fibers will get embedded in the yarn structure without any strong binding. These fibers can be released from the structure in the later stage while wearing or washing the garments [22]. The microfiber shedding can also be compared with the pilling nature of the fabric. The fiber shedding can be considered as the next stage of pill formation which is pill wear-off [23]. The process of pill formation can be simplified into 3 steps [24]:

i. Formation of fuzz
ii. Entanglement of fuzz to form pills
iii. Pill wear-off

In the fuzz formation step, the fibers get protruded from the yarn structure and it mostly happens in the dry state when the material is subjected to wear and use [25]. Figure 2 represents fiber protrusion on the fabric surface. These protruded fibers can get entangled to form pills on the surface of the fabrics which is held due to the tenacity of the fibers and then they wear-off from the surface [26]. Those pills which are very small in size can be added on to the category of microfibers. Figure 3 represents the predicted shedding mechanism of microfibers from synthetic textiles.

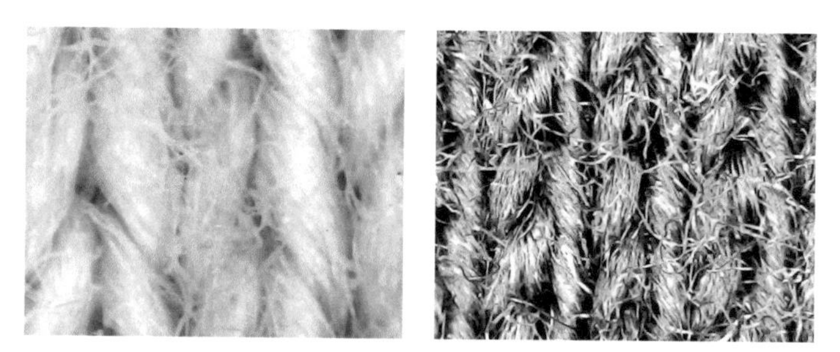

Fig. 2 Microscopic images of fiber protrusion on the fabric surface [Authors own representation]

However, the formation of pills is not necessarily important as a preceding step for the microfiber shedding to happen. Moreover, the shedding property depends mostly on the fuzz formation step where the short, loose, or damaged fibers protrude from the surface. These protruded fibers can directly disentangle from the surface even before the formation of pills. Hence, the fuzz forming nature of fibers determines the level of shedding. The more fuzz formation causes increased shedding [23]. In general, all the types of materials including natural and synthetic textiles shed fibers [23] during all the stages of their life cycle [28]. But the amount or rate of shedding can vary with various factors. It can vary with the characteristics of the textile material like fiber type [23, 27], yarn type [23], fabric construction [29] as well as the external factors like mechanical and chemical actions during washing [23, 27] and wearing [12]. Figure 4 represents the microscopic image of the microfiber released from synthetic apparel.

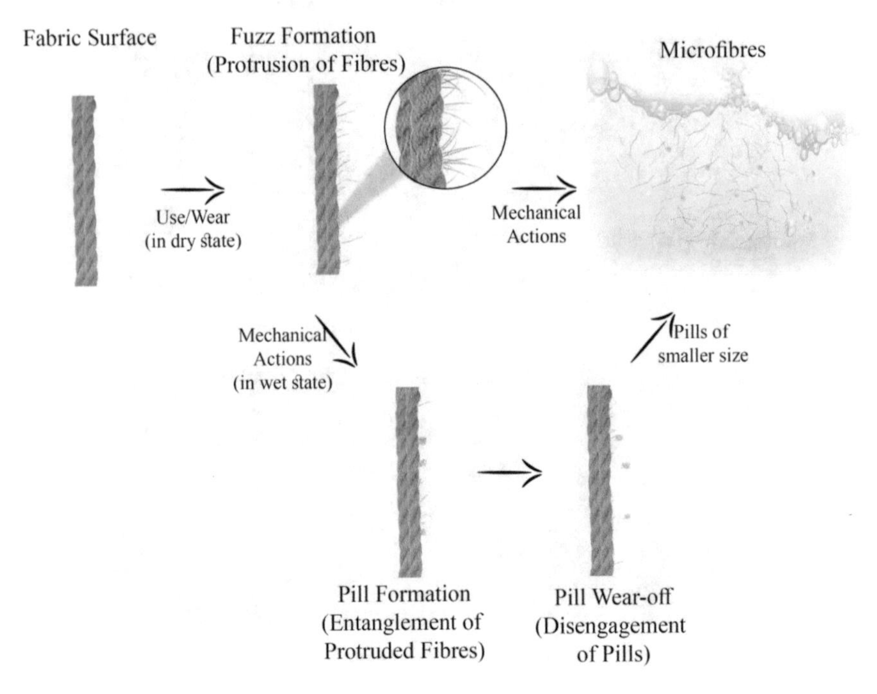

Fig. 3 Microfiber shedding mechanism as predicted by the previous literature [Authors own representation]

3 Microfiber Generations in Different Phases of Life Cycle

Since the textile materials are subjected to considerable mechanical and chemical actions that are responsible for the shedding throughout their life cycle, the microfiber shedding can occur during all the stages including production, consumption, and disposal stages [30].

3.1 Production Stage

In the production stage of textile materials, wet processing plays a vital role in the microfiber emission. The wet processing of textiles is one of the important processes that cannot be neglected as it is responsible for the improved value of the materials. The aesthetics, comfort, and other functional properties of the textiles are potentially improved in wet processing. Throughout the process, a huge variety of chemicals are used in the form of dyes, finishing agents, and other auxiliaries [31]. The fabrics

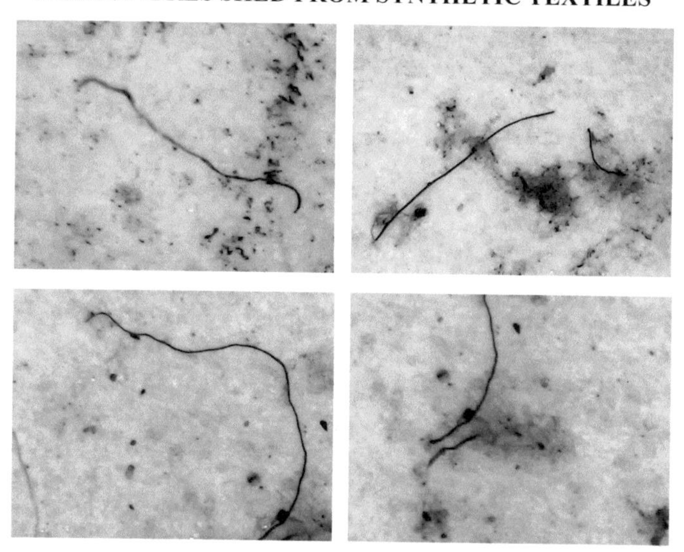

Fig. 4 Microscopic images of microfibers shed from synthetic apparels during laundry [Authors own representation]

are subjected to more forces, both mechanical and chemical actions, during the dyeing and printing process than during the domestic laundry and this can cause microfiber shedding. The release of an enormous amount of effluents from the wet processing industries can easily carry away the microfibers and eases the release into water systems. Hongjie Zhou et al. have investigated the presence of microfibers in the textile printing and dyeing effluents that are properly treated with the wastewater treatment system. They have discovered that almost 85–99% of microfibers are eliminated in the wastewater treatment. Yet the number of microfibers found in textile wastewater is significantly high. With this higher level of microfibers in dyeing and printing effluents, they have concluded that wastewater from the textile production plant contributes majorly to the presence of microfibers in the natural water bodies [20]. The other researchers have analyzed the Waste Water Treatment Plant (WWTP) of the textile industry with 30,000 tons of treatment capacity in which printing and dyeing effluent accounts for 95% of wastewater. Their study aims to understand the effectiveness of different stages of WWTP in the removal of microfibers in the effluent. They have collected samples from each stage of the treatment and characterized the microparticles found. It has been noted that microfibers were dominant and they account for 80-100% of the microparticles found in the sample sites. They have also concluded that both natural and synthetic fibers are found in the samples out of which microplastic fibers (synthetic fibers) account for 60% of total fibers [32]. The presence of a huge number of microfibers in the effluents of textile production plants confirms that the shedding occurs in the production stage of textile materials and microfibers are being emitted into the environment particularly into water systems.

The higher number of microfibers in the effluents of dyeing and printing industries than those found in the municipal sewage plants [20] shows the dominant role of the production process of textiles in the microfiber shedding.

3.2 Consumption Stage

In the consumption stage of textile materials, wearing and maintenance of garments and apparels are common. As a consequence of wearing apparel, the microfibers are directly released into the atmosphere. While wearing the garments, they are subjected to considerable stress and abrasion that can cause damage to the textile materials. Researchers have studied the release of microfibers into the air as the result of wearing the garments. They have reported that the emission of microfibers due to wearing varies concerning various textile parameters [12]. Another study has been made to analyze the microfibers in the atmosphere by sampling outdoors at different time frames. They have reported that the quantity of microfibers in the atmosphere varies with various factors including consumption habits, socioeconomic status, traffic, and urbanization. They have also noticed a significant variation in the fibers including cotton, wool, acrylic, polyester, and polyamides at different time frames. With that, they have concluded that the variation in the clothing requirements in different seasons can attribute to the varied fibers in the atmosphere [16].

Washing is the most common maintenance done for apparel. Washing is usually carried out during the consumption stage to clean and remove dirt and stains. Domestic laundering is considered to be the major source of microplastic fibers in the environment [28]. Generally, in the washing process, the fabrics are saturated with water and then agitation is done in the presence of additives that can aid the dirt removal. Then, the fabrics will be squeezed and dried. In the viewpoint of textile technologists, it is very similar to the mild scouring process done in the wet processing of textiles [33]. During washing, the fabrics are subjected to considerable mechanical (agitation) and chemical actions (detergents and other additives). About 90% of damage to the textiles is caused during washing than wearing which leads to shedding [28]. Domestic laundering accounts for 34.8% of microplastics in the environment [4]. A report has estimated that Australia which is having less than 1% of total washing machines in the world could release nearly 62 kgs of microfibers per week [28]. Around 18,000,000 microfibers (synthetic fibers) can be released as an average from a domestic wash load of 6 kg of synthetic textiles [34]. Since domestic laundry contributes more to the microfiber pollution, various studies have been made in the essence of analyzing the different washing parameters like washing machine type [29], washing duration [22], washing temperature [23], water volume [35], and additives like detergent and softeners [21, 28] on the number of microfibers that are shedding from the textiles and apparels during laundry. Figure 5 represents the life cycle of microplastic from apparel and textiles.

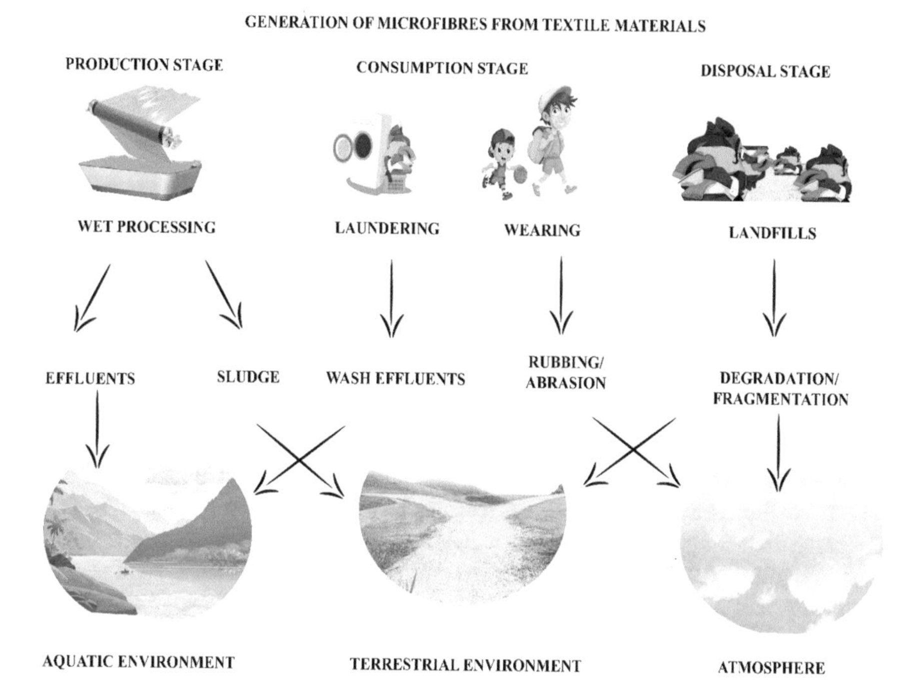

Fig. 5 Sources and life cycle of microfibers from textiles and apparels [Authors own representation]

3.3 *Disposal Stage*

In the disposal stage, textile materials contribute to the microfiber pollution as secondary microplastics which are generated as the result of the breakdown of larger plastics into smaller fragments [4]. Fast fashion accelerates the fashion cycle and leads to overconsumption and higher disposal. The rapid changes in the fashion trends make the consumers buy new clothes frequently and their lesser price encourages the consumers to dispose them easily without any second thought. The utilization of the garments gets decreased as a result of fast fashion which imparts rapid changes in fashion trends. The usage period of clothes gets reduced by 36% in the past 15 years [8]. After consumption, the apparels are disposed and they end up in landfills. In that stage, the microfiber shedding occurs due to the breakdown of textile materials [30]. The degradation and fragmentation of synthetic textiles can release microfibers into the environment [36]. The synthetic textiles slowly degrade over long periods and release microfibers and nanofibers. These microfibers and nanofibers can end up in the atmosphere, water resources by leaching, or get deposited on the land sites. The release of microfibers from the textiles on landfills could be the potential source of microfibers in the terrestrial environment [37]. However, there is no detailed research on the shedding behavior of textiles in their disposal stage due to degradation over a longer period. The textile materials can shed microfibers in all the stages of their life cycle due to varied mechanical and chemical actions to which they are subjected to.

In each stage, the nature of external actions that accelerates the shedding can vary. However, the common thing in all the stages could be the textile materials. Though the external factors accelerate the shedding, the role of textile parameters, and their characteristics in causing shedding cannot be neglected.

4 Effect of Textile Parameters in Microfiber Shedding

The inherent properties of textiles materials that differ with the various parameters contribute to the dissimilarities in microfiber shedding from one material to another. Under similar conditions, the shedding of one material is not similar to the other and this shows the prominent role which is played by the textile parameters. The ability of the textile materials to respond to the external factors varies with the textile characteristics which include the fiber, yarn, fabric, and other characteristics. The various textile parameters that can affect the level of shedding includes the type of fibers, fiber length, yarn type, yarn twist, yarn hairiness, yarn strength, fabric structure, pilling resistance, abrasion resistance properties of fabrics, type of surface treatments, and fabric finishing.

4.1 Effect of Fiber Properties

All types of fibers tend to cause shedding; however, the quantity of shedding varies with the nature of different fibers. Various studies have been done to understand the effect of fiber type and their characteristics which can potentially influence the microfiber shedding.

4.1.1 Type/Nature of Fiber

In apparels, a wide variety of fibers including natural, synthetic, and regenerated fibers are being used. Among all those fibers, cotton (natural) and polyester (synthetic) fibers are more dominant which holds around 26.05 and 51.5% of total fiber production respectively [38]. A considerable level of variation in the shedding can be seen depending on the nature or origin of the fiber. To support this variation concerning the fiber type, various researchers have studied the shedding behavior of different fibers and the effect of fiber type on shedding.

Napper et al. studied the impact of fiber type on the shedding by comparing the shedding behavior of 100% acrylic, 100% polyester, and 65/35 P/C blend. They have reported that there was higher shedding in the case of 100% acrylic and 100% polyester while comparing with the P/C blend. They have added that P/C blends shed 80% fewer fibers than acrylic [27]. In contrast to this, the other researchers have reported that the cellulose-based fibers shed more than the synthetic fibers.

Zambrano et al. studied the impact of fiber type by varying the fiber type without changing the fabric structure so that the effect of fiber type on shedding can be analyzed effectively. They have studied the shedding behavior of 100% cotton, 100% rayon, 100% polyester, and 50/50 polyester/cotton blend fabrics with common fabric structures (Interlock). They have reported a significant level of shedding in all fiber types and a higher level of shedding is noted with cotton, rayon, and P/C blends than the polyester fibers. They have concluded that the cellulose-based fibers shed more fibers than synthetic fibers. They have added that the varied fuzz formation (protrusion of fibers) nature of fibers could be the reason for the variation in the shedding property. It should be noted that the fuzz forming ability of the fiber again depends on the shape, thickness, stiffness of the fibers [23]. LibiaoYang et al. studied the shedding behavior of most commonly used synthetic fabrics—polyester and polyamide and a man-made fiber—acetate. Among these, higher shedding is noted in the case of acetate fibers even though all the fibers shed significantly higher [29]. This again confirms the higher shedding nature of cellulose-based fibers. A similar pattern of the result was obtained in a study that analyzed the shedding behavior of denim fabrics made of 100% cotton, 100% polyester, and P/C blend where higher shedding is noted for cotton followed by P/C blend and then by 100% polyester [39].

A study made to analyze the real-time washing load as the domestic laundry plays a dominant role in microfiber emission has reported the presence of natural fibers at a higher level in the collected fibers from a wash load of mixed fabrics. The dominance of natural fibers (cotton, wool, and viscose) was extremely high as it accounts for 96% whereas the synthetic fibers were only 4%. This supports the finding of researchers who reported higher shedding with natural fibers in the real-time situation [40]. Cesa et al. who studied the 100% cotton, 100% polyester, 100% acrylic, and 100% polyamide materials in garment form for microfiber shedding behavior have noticed higher shedding in the cotton followed by acrylic and a similar level of shedding for polyester and polyamide. They have concluded that fiber length can contribute to the shedding and cotton being shorter fibers shed more than other fibers. The hydrophilic nature of cotton could also be the potential reason for increased shedding and also its lower tenacity when compared to polyester and polyamide increases the shedding [26]. Next to cotton, a higher level of shedding is noted in acrylic among all the synthetic fibers. The lower tenacity of acrylic fibers accounts for the higher shedding. Here, it has to be noted that when cotton and acrylic are compared, acrylic has the least tenacity yet shedding is more in the case of cotton. It is attributed to the shorter fiber length of cotton. Hence, it is clear that the factor affecting the shedding is not limited to an individual characteristic of fiber rather it depends on the combination of characteristics [26].

The emerging dominance of recycled fibers due to their environmental and economic benefits urged the study of shedding property of recycled fibers. The recycled fibers can be easily broken down due to the reduced molecular weight and molecular chain length and they exhibit poor mechanical properties. Their poor mechanical properties due to the structural changes resulted from the recycling process. The thermal exposure and shear degradation of the polymers during the recycling process is responsible for these structural changes. Moreover, the surface of the recycled fibers seems to be duller and more deformed. These changes (strength reduction and deformed surface) in the recycled fiber characteristics cause increased shedding. The researchers analyzed and compared the shedding behavior of virgin and recycled polyester fibers and found 2.3 times more shedding in the case of recycled polyesters than virgin polyester [41].

In contrast, other researchers noticed lesser shedding in the case of recycled fibers. Frost et al. studied the impact of recycled fiber contents on the shedding behavior of the fabrics. They analyzed the shedding behavior of 100% virgin cotton, 20% recycled cotton, 40% recycled cotton, 100% virgin polyester, 70% recycled polyester, and 40% recycled polyester. In their study, they have found no significant difference between the shedding of virgin and recycled cotton. They have reported that 70% recycled polyester shed fewer fibers than the 40% recycled fibers. Further, results showed that 70% recycled polyester sheds longer fiber than the other polyester fibers. The reason behind this could be the higher elongation of recycled polyester at the breaking point. Hence, they have concluded that the recycled fibers showed lesser shedding as they tend to elongate rather than break [42]. The same result has been obtained in the other study where virgin and recycled polyester fabrics are tested for shedding. The researchers have reported higher shedding in the case of virgin polyester than the recycled polyester [43]. Table 1 consolidates the previous analysis performed with different fiber types.

4.1.2 Fiber Length (Staple/Filament)

Staple fibers and filament types are the two configurations of fibers. The staple fibers are of shorter length whereas the filaments are of indefinite length [46]. The slippage of staple fibers from the yarn structure is easy because of its shorter length. During the spinning process, the twists are inserted to keep the fibers together. It is common that during the twist insertion, fibers of shorter length slides toward the sheath and the longer fibers move toward the core. Since the short fibers are on the sheath it is easier to get disengaged from the yarn structure [41]. In the case of staple fibers, even though the material is not damaged due to external actions, there will be a release of shorter fibers [22]. However, in the filaments, the shedding happens only after the breakdown of filaments due to external forces. This attributes for higher shedding in the case of staple fibers when compared to filaments [41]. The researchers have studied the shedding property of polyester knit, polyester woven, and polypropylene woven and reported lesser shedding in the polyester knit sample than the other two. This is attributed to the filament nature of fibers used in the polyester knit whereas polyester and polypropylene woven fabrics were made of staple fibers [45]. The other researchers who have noticed higher shedding in the case of cotton than the polyester have concluded that it can be attributed to the staple nature of cotton whereas the polyester fibers are filaments [23]. Yang et al. have also claimed that the lesser shedding of polyester than that of acetate and polyamide is due to the continuous filament nature of polyester [29]. Francesca et al. have noticed

Table 1 Fibers parameters that affect shedding as reported in the literature

S.no	Fibers studied	Shedding order (Highest to Lowest)	Noted driving factor	References
1.	• 100% Acrylic • 100% Polyester • 65/35 P/C blend	i. Acrylic ii. Polyester iii. P/C Blend	–	[42]
2.	• 100% Cotton • 100% Rayon • 100% Polyester • 50/50 P/C blend	i. Cotton ii. Rayon iii. P/C Blend iv. Polyester	Fuzz forming nature of fibers accelerates shedding Fiber length (Staple fibers shed more)	[23]
3.	• Polyester • Polyamide • Acetate	i. Acetate ii. Polyamide iii. Polyester	Fiber length (Staple fibers shed more)	[29]
4.	• 100% Cotton • 100% Polyester • 100% Acrylic • 100% Polyamide	i. Cotton ii. Acrylic iii. Polyamide, Polyester	Fiber tenacity and fiber length (Lower tenacity and shorter length increases shedding)	[26]
5.	• 100% Cotton • 100% Polyester • 40/60 P/C blend	i. Cotton ii. P/C Blend iii. Polyester	–	[39]
6.	• Recycled polyester • Virgin polyester	i. Recycled Polyester ii. Virgin Polyester	Fiber strength (Reduced Strength of recycled fibers increases shedding) Fiber length (Shorter fibers shed more)	[41]
7.	• 100% Virgin polyester • 70% Recycled polyester • 40% Recycled polyester	i. 40% Recycled Polyester ii. 100% Virgin Polyester iii. 70% Recycled polyester	Fiber elongation (Higher elongation causes lesser breakage)	[42]
8.	• 100% Polyester • 65% Recycled polyester • 100% Polyester, 50/50 Cotton/Modal combination	i. 100% Polyester, 50/50 Cotton/Modal combination ii. 100% Polyester iii. 65% Recycled polyester	Fiber length (Staple fibers shed more)	[44]
9.	• Polyester (Staple) • Polyester (Filament) • Polypropylene (Staple)	i. Polyester (Staple) ii. Polypropylene (Staple) iii. Polyester (Filament)	Fiber length (Staple fibers shed more)	[45]

(continued)

Table 1 (continued)

S.no	Fibers studied	Shedding order (Highest to Lowest)	Noted driving factor	References
10.	• Real-time wash load	i. Natural fibers (Cotton, Wool, Viscose) ii. Synthetic fibers (Polyester, Acrylic, Nylon)	Origin of fibers (Natural fibers shed more than synthetic fibers)	[40]

higher shedding in the case of double structure fabric (modal and cotton blend in the backside and 100% polyester in the front side) than the fabrics made of 100% polyester. They have reported that the staple fibers on the backside of the double structure fabric could be the reason for higher shedding. They have supported their conclusion by examining the shed fibers where they found more cellulosic fibers which form the backside of the structure than the polyester fibers on the front side [44]. The same conclusion was provided by the other researchers who studied the shedding of polyester fabrics which are made of staple and filament yarns. They have also noted that fabrics which are made of staple fibers shed more than the fabrics made of filament fibers [21]. In contrast to all these, the other researchers who quantified the microfiber shedding by the sonication extraction method have reported that there is no significant difference in microfiber shedding between the yarns made of filament and staple fibers [47]. Figure 6 represents the various fiber parameter that influences the microfiber shedding.

To consolidate, the effect of fiber properties and characteristics on the microfiber shedding is inevitable. They can directly or indirectly affect the shedding nature of the fabric. Various researchers have reported that physical and mechanical characteristics of fibers like fiber length, fiber tenacity, fiber elongation can affect or change the microfiber shedding behavior. Among the parameters discussed, fiber length plays a dominant role. The staple fibers shed more than the filaments as short fibers can easily slip away from the structure [23, 26, 29, 44, 45]. Since the natural fibers are of staple length, more shedding is noted with natural fibers. Along with that the researcher also mentioned that the fibers with cellulosic content shed more fibers [23, 26, 29, 40]. The fiber tenacity and fiber strength could also contribute to shedding. The reduced tenacity (for fiber-like acrylic) can lead to easy breakage and cause more shedding. This causes increased shedding in the case of recycled fibers where the recycling process can reduce the mechanical properties of the fibers [41]. It is also worth to mention the biodegradable nature of the natural fiber is their main advantage and so less concern was given to their microfiber releasing capacity. But in the case of synthetic fibers like acrylic and polyester, due to their potential nonbiodegradability, they possess a higher threat to the environment.

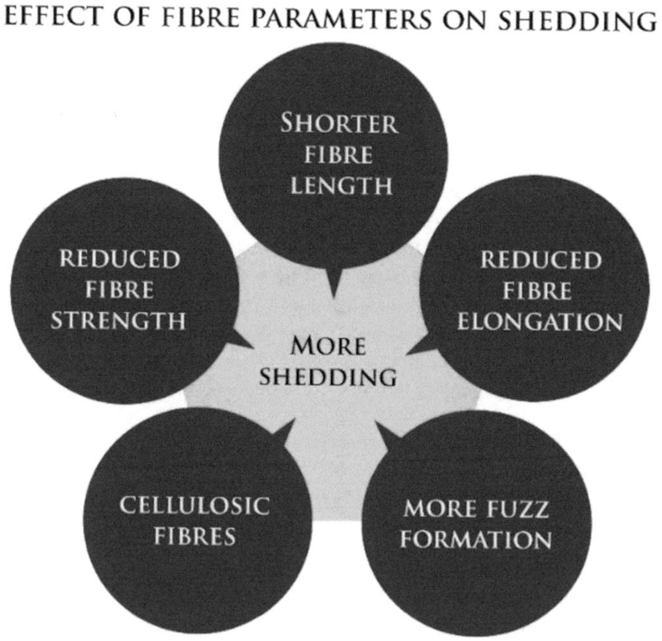

Fig. 6 Effect of fiber properties and characteristics on shedding [Authors own representation]

4.2 *Effect of Yarn Parameters*

Yarns are made by assembling the fibers or filaments with or without twist [46]. The properties of yarn including hairiness, breaking strength, and evenness can be varied with the fibers used and the spinning methods adapted. These differed properties can have a potential impact on microfiber shedding [23].

4.2.1 Type of Yarn

The characteristic of yarn hugely depends on spinning methods. Various spinning methods including rotor spinning, ring spinning, air-jet spinning, compact spinning, self-twist spinning, and hollow spindle spinning will produce different yarn structure [48]. Cai et al. studied the microfiber shedding from different yarn types such as air-jet yarn, ring-spun yarn, and rotor yarn. They have reported a higher level of shedding in the case of rotor yarns when compared with air-jet and ring-spun yarns. In the ring spinning and air-jet spinning processes, the short fibers are removed as the fibers are drawn and parallelized [47]. While comparing conventional ring-spun yarn with yarns produced by the compact ring spinning method, the hairiness will be higher in the case of the conventional method and this can cause increased fiber shedding [41].

4.2.2 Tensile Properties of Yarn

The tensile property of the yarn is very significant in the case of microfiber shedding. As the yarn twist, count, and fiber staple length are interrelated in the case of tensile strength, a lower tensile strength may allow the fiber to break and shed during continuous limited pressure applications like wearing and washing of textiles. As the shedding mechanism is connected with several factors, it is very difficult to correlate with any particular properties as no such studies exist. However, the tensile strength of fiber and yarn is important for fiber fracture and flex fatigue. During laundry, at the wet condition, the tensile strength of the cellulose-based yarns are noted poor whereas, the synthetics like polyester showed comparatively good strength.

4.2.3 Hairiness and Evenness

While determining the surface properties of the textiles, yarn hairiness is an unavoidable parameter. Hairiness is the protrusion of fibers from the surface structure [41]. The hairiness and evenness of the materials have shown an impact on the shedding. The hairiness and evenness of yarn interdependently affect the shedding. Less hairiness and more evenness can reduce shedding [23]. The statistical analysis done by the other researcher showed a significant level of increase in the fiber shedding in the case of fabrics having hairiness of 4 mm and longer [41]. The comparison of shedding of polyester woven, polyester knit, and polypropylene woven, showed a reduced shedding in the case of the knit structure where there is no hairiness. But while comparing the other two fabrics, the hairiness level is the same yet their degree of shedding varied. Hence it has to be noted that hairiness can be one of the factors and not the individual factor that determines the shedding [45]. The increased yarn hairiness can have benefits of improved softness and thermal insulation whereas when comes to microfiber shedding, the increase in the hairiness leads to increased shedding [41].

4.2.4 Yarn Count and Twist

Yarn count indicates the linear density of the yarn [46]. Whenever the yarn count gets increased, the number of fibers per unit cross-section will increase. This leads to more microfiber shedding as more fibers are there in the yarn structure [29]. Francesca et al. have reported that the fabrics made of yarns with a higher twist shed lesser fibers than the fabrics made of yarns with a low twist or no twist [44]. When the yarn gets more twisted, it will get a tighter structure and this can reduce the shedding [47].

From the available research data, it can be noted that the higher tensile strength of the yarn reduces the microfiber shedding as it required more force to break the fiber from its surface. Similarly, fabrics with low hairiness and high evenness shed fewer fibers during laundry. The effect of the twist was also noted significant as loose yarn tends to release more fiber, a higher twist factor is preferred for lower microfiber shedding.

4.3 *Effect of Fabric Parameters and Properties*

Fabric parameters which include the structure [29], tightness factor [29], and thickness [45] also have a considerable impact on the shedding nature of the textile material.

4.3.1 Structure and Tightness/Cover Factor

The conversion of yarn into fabrics can be made by using three principles: interweaving, intertwining, and inter looping. Among these, interweaving (woven) and inter looping (knitted) are the most commonly used techniques in the case of textile products [46]. Generally, the knitted fabrics are less compact while comparing with woven fabrics due to their way of construction. Researchers reported that the knit structures being less compact tend to shed more while compared with the woven structure [29]. When there are a greater number of yarns in the unit area of fabric, the structure gets tighter. This tighter structure can withstand the abrasive forces and thus leads to lesser shedding [29]. In contrast to this, Almroth et al. have reported that the tighter structure sheds more fibers than the loose structure. They have supported their finding by the fact that tighter structure allows more fibers in the unit area of the fabric which results in increased surface abrasion and so the microfiber shedding. Their findings showed a double fold increment in the shedding [21]. Other studies showed higher microfiber shedding with woven fabric due to the types of yarn and its hairiness compared to a knitted fabric. The microscopical analysis revealed that the role of fiber length might contribute to the shedding of the microfiber. The knitted fabric is made of filament yarn and the warp, weft yarns of the woven fabric are made of staple fibers. This attributes higher hairiness in the woven fabric than knitted fabric. Hence, the researcher hypothesized that staple fibers will easily slip away from the yarn and leads to higher microfiber release in the laundry [45].

Similarly, higher Grams per square meter (GSM) or thickness may result in increased fabric weight. This can increase the number of fibers in the unit area and the increased number of fibers in the unit area can lead to increased shedding. However, this cannot be directly related to shedding. The other researcher has found reduced shedding in the case of fabric with higher GSM [45].

4.3.2 Effect of Fabric Properties

The properties of fabrics such as pilling and abrasion resistance can also contribute to the shedding behavior of the fabric. Researchers have claimed that these properties have a considerable impact on microfiber shedding [23, 27]. As discussed in Sect. 2, pilling is another important parameter that has a direct influence on the microfiber shedding. Pilling is a phenomenon where the protruding fibers get entangled to form fiber balls (also known as pills) on the fabric surface. Napper et al. who studied the variation in the amount of shedding concerning fabric type (100% acrylic, 100% polyester, 65/35 P/C Blend) have reported that synthetic fibers shed more than that of natural and synthetic blends. They have also reported that the pilling nature of synthetic materials contributes to the higher shedding [27]. Generally, polyester pills more. The reason behind this is the higher tenacity of polyester fibers which holds the pills on the surface and rarely releases the fibers. One of the ways of reducing pills on the surface is to reduce the tenacity of fibers that can ultimately lead to pill break off due to fiber fatigue and improves the surface appearance [49]. This is generally done in the fiber production stage itself where the additives like polyacid or polyhydric alcohol and branching agents like pentaerythritol are added to the polymer itself [50]. But the wear-off of pills can lead to an increase in the fiber shedding [27]. Yang et al. who noticed higher shedding in the acetate fabric compared with polyamide and polyester fabrics. They have also noticed that acetate fabric has exhibited poor pilling resistance than the other two fabrics. With this, researchers concluded that the fabric with poor pilling resistance tends to cause more shedding [23, 29]. It is important to note that pilling can contribute to shedding to some extend but pill formation is not essential as a previous step for shedding to happen [26].

4.3.3 Abrasion Resistance

Abrasion Resistance is the ability of the material to withstand the physical destruction of fibers, yarns, and fabrics which results from the relative motion of the textile surface over another surface during wear of the materials [51]. A textile material's abrasion resistance can depend on various factors including fiber fineness, yarn count, yarn type, fabric construction, etc. [52]. It has been noted that the loss in weight or reduction in fabric thickness due to abrasion is expressed as the elimination of fibers of shorter length [53]. And hence the microfiber shedding can be compared with the abrasion resistance property of the fabric. Zambrano et al. have analyzed the abrasion resistance properties of the fabrics to compare them with the microfiber shedding characteristics. They have concluded that the fabric with higher abrasion resistance can withstand the stress during the laundry and exhibit lesser shedding [23].

4.4 Effect of Surface Finishes

Surface treatments or surface finishing is an important process of textile production. Surface finishes are done to improve the properties of the textile materials or to impart desired properties. Fabrics can be subjected to a wide variety of mechanical and chemical treatments to achieve the required properties. However, these surface treatments can potentially affect microfiber shedding behavior. Various studies have been made to understand the variation in shedding behavior for different finishing techniques.

4.4.1 Mechanical Finishes

Mechanical finishes like singeing, calendaring, raising, and brushing are usually done to improve the handle and alter the bulkiness of the fabrics. The shedding behavior can be varied with different finishes. In the raising process, the surface of the material is brushed to improve bulkiness and softness. As a result, fibers will be protruding on the surface which can easily slip from the surface and this can probably increase the shedding. However, in singeing and calendaring, the protruding fibers are eliminated and this can cause reduced shedding [54]. Cai et al. studied the effects of mechanical surface treatments on the shedding behavior of textile materials. In their study, they analyzed the fleece and plain brushed fabrics and compared them with the fabrics without surface treatments. In the case of fleece fabrics, the surface fibers are cut by blades in the shearing process to get the desired surface effect whereas, in the case of plain brushed fabrics, metal brushes are used to break the surface fibers to produce the desired effect. They have noticed a significant impact of surface treatment on the shedding. They have reported that the fabrics which have undergone mechanical surface treatments have shed a higher number of microfibers than the untreated fabrics. They have supported their findings with the fact that surface treatments like shearing and brushing cause the generation of loose surface fibers and also loosen the yarn and surface structures, thereby increasing the shedding [55].

The other researchers have also reported the same conclusion that the brushing of fabrics has a significant impact on the shedding. A variety of fabric surfaces can be obtained with a wide variety of brushing techniques. Though the effect of different brushing techniques is not studied, it has to be noted that the shedding can be reduced with reduced brushing [43]. In the other research where the different method of microfiber extraction (ultrasonication) is done, the same result was obtained. They have also noticed higher shedding in the case of samples with surface treatments (fleece and brushed surface) than the untreated sample [47].

4.4.2 Chemical Finishes

Generally, chemical finishes are not responsible for microfiber generation. Usually, in chemical finishing, the chemicals bind the protrusion in the surface and form a layer over the fabric surface. Since the protruding fibers are bound within the surface, the fiber release will be lesser. Certain finishes can be made to reduce shedding [54]. Some researches have been made to understand the effectiveness of chemical finishes on reducing the microfiber shedding. De Falco Francesca et al. have studied the effect of pectin-based finishes in reducing the microfiber shedding of woven polyamide fabrics by padding process. They have provided pectin modified with glycidyl methacrylate coating on the surface of polyamide fabrics with different concentrations. The best performance was noted where pectin and glycidyl methacrylate were in the ratio of 1:1. They have reported a 90% reduction in the microfibers released in the case of treated fabrics when compared with the untreated fabrics. The SEM analysis of treated polyamide fabric has shown that the bio-based coating has reduced the fragmentation of the polyamide fabrics during the washing. It is also reported that the treatment can withstand the washing by post-wash analysis [56]. In other research, biodegradable polymers Poly Lactic Acid (PLA) and Poly Butylene succinate-co-butylene adipate (PBSA) are coated on polyamide fabrics to reduce microfiber shedding. In this, electrofluidodynamic process is carried out to provide better homogeneity of coating while not affecting the handle and wettability of the fabric. They have reported a strong reduction, more than 80%, in the microfiber generation of treated fabrics compared to untreated ones [57].

To consolidate, the surface treatments have a significant impact on the microfiber shedding. Mechanical finishes have accelerated the shedding [43, 47, 55] whereas the chemical finishes have controlled the shedding to a considerable extent [56, 57]. However, only the effect of brushing is studied among various mechanical finishes. The other finishes like calendaring and singeing might reduce the shedding where the detailed researches are not found. In the case of chemical finishes, the effective reduction has been noted. Yet the researches have been done on polyamide fabrics alone. The other point to be considered with chemical finishes is durability as the finishes will degrade over the domestic laundering of fabrics.

4.5 *Effects of Aging*

The time elapsed since the material was formed is referred to as the age of the material. The aging can be of different types including physical aging, thermal degradation, chemical attack, mechanical stress, and photochemical degradation. Throughout the material usage, it undergoes different types of aging and the effect of aging on the textile properties will vary with types of aging. As a result of physical aging, the fibers get harder, denser, and stiffer. Further, the exposure of the textile materials to light throughout the usage can cause photochemical degradation which can result in the reduction of molecular weight of the polymer [58].

As a result of aging, there will be considerable changes in the physical and mechanical characteristics of the textile materials. These changes can affect the shedding behavior of textile materials. Hence it is important to analyze the effect of garment aging on the shedding behavior of the textile materials. Various studies have been made to understand the effect of aging on the microfiber shedding. Hartline et al. studied the microfiber shedding of new and aged garments. They simulated the aged garments by subjecting the fabrics to 24 h of continuous washing. The mechanically aged garments shown increased shedding and shed 25% more fibers than the new garments. They have used a filtering system where two different pore sizes were used. The significant difference between the new and aged garments was noted in the 333 μm filter and no significant variation is noted in 20 μm filter. From this, they have concluded that the aging of garments causes an impact on the larger fibers than the shorter fibers. Moreover, more fraying was noted in the visual examination of aged garments and this can attribute to the increased shedding in the aged garments [59]. Similarly, the other researchers have analyzed the impact of aging on shedding by simulating aged effect by repolishing. They have noted higher shedding for repolished fabrics than the non-repolished fabrics indicating higher shedding for used or worn garments. The repolishing of fabrics can affect the surface and cause damage to the yarns. The damaged yarns with cut edges increase the fiber release [9].

Bruce et al. have analyzed the shedding of new and aged jackets. The statistical analysis of the fiber mass released from the aged and new jackets shown a significant difference. They have noticed 60% more fibers from aged jackets. They have also suggested that the reason for higher shedding in the aged garments might be the weakened fibers which are resulted out of wear. Moreover, they have also noticed that the fibers shed from the aged garments were larger than the fibers from new jackets. They have also added that the aging nature of the fiber will differ from one to another and the studies should be made to assess the effect of aging of different materials like nylon, acrylic, rayon, and others on the microfiber shedding [60]. The other researchers studied the effect of aging effect on shedding by analyzing different fabrics including polyester knit, polyester fleece, nylon, and acrylic fabrics. For the simulation of aging, repolishing is done. As similar to previous researches, higher shedding is noted in the repolished fabrics than the untreated ones indicating higher shedding with aging. They have noted that in the case of polyester knits, with aging, the fibers released were longer whereas in the case of nylon and acrylic the length of the fibers was small [21]. The other researcher has studied the effect of aging by comparing the fibers shed from 1st to 5th washes with that of 11th to 15th washes. The higher level of microfiber release is noted in 11th to 15th washes (0.0046% w/w) and it is almost twice the quantity of fibers released in the initial washes (0.0024% w/w) [28]. The study made by Ganesh Lamichchane to evaluate the shedding behavior of new and old polyester jackets also reported a similar result. The old jackets had shed a greater number of fibers than the new jackets. It has to be noted that the shedding of the old jacket is 1.6 times more than that of new jackets [61]. Figure 7 consolidates the different yarn, fabric and finishing properties on microfiber shedding behaviour.

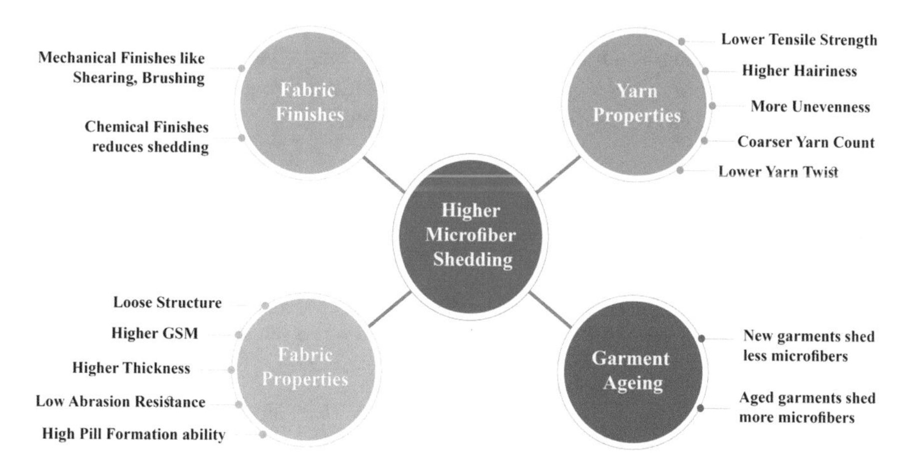

Fig. 7 Properties of yarn and fabric that increases microfiber shedding [Authors own representation]

All the researches which have been made so far have reported a significant difference in the shedding of new and aged garments and noticed higher shedding in the case of aged garments [13, 21, 28, 59–61]. The shedding of aged garments is found to be 25–80% more than the new garments [28]. It is attributed to the weakening of materials due to the changes in the mechanical and chemical characteristics resulted from aging. Moreover, it can be concluded that the stiffer, harder fibers can cause more shedding as the aging can make the fibers stiffer and harder [58]. However, the aging effects are simulated under laboratory conditions. Further, more detailed researches have to be made in order to understand the effect of real-time aging as there might be some significant changes in the shedding pattern.

5 Summary and Recommendations

Microfiber shedding of synthetic textile is mainly attributed to the textile characteristics of the fabrics used. It is estimated that the microfibers shed during the manufacturing, usage, and disposal phase of the textile. Though the laundering process is noted as one of the main processes which release the microfiber, the fabric parameters decide the amount of fiber release. The types of fibers, yarn types, or yarn manufacturing method, and fabric properties have a higher impact on the microfiber shedding. Proper control or engineering of yarn, fabric parameters will help in reducing the microfiber release from the textile. However, very few research works performed in the textile discipline to control the parameter effectively. Majority of the literature works performed from the environmental science region and the results were correlated with the textile parameters using literature. But, no direct measurements were performed still. Hence, it is necessary to perform such research from the textile domain, to understand the interaction between the textile parameters and microfiber release from the textile. As many of the literatures reported the mechanical agitation as the main source of microfiber generation, it is important to correlate the parameters like tensile strength, abrasion resistance, and pilling characteristics. A specific research by controlling these parameters will provide the real correlation between microfiber release and the abovementioned parameters.

References

1. Plastic Market Size, Share & Trends Analysis Report By Product (PE, PP, PU, PVC, PET, Polystyrene, ABS, PBT, PPO, Epoxy Polymers, LCP,PC, Polyamide), By Application, By Region, And Segment Forecasts, 2020–2027. https://www.grandviewresearch.com/industry-analysis/global-plastics-market. Accessed 28 Aug 2020
2. Browne MA, Galloway T, Thompson R (2007) Microplastic—an emerging contaminant of potential concern? Integr Environ Assess Manag 3(4):559–561
3. Cole M, Lindeque P, Halsband C, Galloway TS (2011) Microplastics as contaminants in the marine environment: a review. Mar Pollut Bull 62:2588–2597. https://doi.org/10.1016/j.marpolbul.2011.09.025

4. Boucher J, Friot D (2017) Primary Microplastics in the oceans: a global evaluation of sources. IUCN, Gland, Switzerland, p 43. https://doi.org/10.2305/IUCN.CH.2017.01.en
5. Young S (2019) The real cost of your clothes: These are the fabrics with the best and worst environmental impact. https://www.independent.co.uk/life-style/fashion/fabrics-environment-fast-fashion-eco-friendly-pollution-waste-polyester-cotton-fur-recycle-a8963921.html. Accessed 28 Aug 2020
6. Liu J, Yang Y, Ding J, Zhu B, Gao W (2019) Microfibers: a preliminary discussion on their definition and sources. Environ Sci Pollut Res 26:29497–29501. https://doi.org/10.1007/s11356-019-06265-w
7. Leonas KK (2018) Textile and apparel industry addresses emerging issue of microfiber pollution. J Textile Apparel Technol Manag 10(4)
8. Ellen MacArthur Foundation (2017) A new textiles economy: redesigning fashion's future. http://www.ellenmacarthurfoundation.org/publications. Accessed 15 April 2020
9. Astrom L (2016) Shedding of synthetic microfibers from textiles, Gothenburg University. https://bioenv.gu.se/digitalAssets/1568/1568686_linn—str–m.pdf. Accessed 15 April 2020
10. Millera RZ, Wattsb AJR, Winslowa BO, Galloway TS, Barrows APW (2017) Mountains to the sea: river study of plastic and non-plastic microfiberpollution in the northeast USA. Mar Pollut Bull. https://doi.org/10.1016/j.marpolbul.2017.07.028
11. Obbard RW, Sadri S, Wong YQ, Khitun AA, Baker I, Thompson RC (2014) Global warming releases microplastic legacy frozen in arctic seaice. Earth's Future 2:315–320. http://dx.doi.org/10.1002/2014EF000240
12. De Falco F, Cocca MC, Avella M, Thompson RC (2020) Microfibre release to water, via laundering, and to air, via everyday use: a comparison between polyester clothing with differing textile parameters. Environ Sci Technol. http://dx.doi.org/10.1021/acs.est.9b06892
13. Browne MA, Crump P, Niven SJ, Teuten E, Tonkin A, Galloway T, Thompson R (2011) Accumulation of microplastic on shorelines woldwide: sources and sinks. Environ Sci Technol 45:9175–9179. https://dx.doi.org/10.1021/es201811s
14. Lahens L, Strady E, Kieu-Le T-C, Dris R, KadaBoukerma ER, Gasperi J, Tassin B (2018) Macroplastic and microplastic contamination assessment of a tropicalriver (Saigon River, Vietnam) transversed by a developing megacity. Environ Pollut 236:661–671
15. Murphy F, Ewins C, Carbonnier F, Quin B (2016) Wastewater treatment works (WwTW) as a source of microplastics inthe aquatic environment. Environ Sci Technol. http://dx.doi.org/10.1021/acs.est.5b05416
16. Kaya AT, Yurtsevera M, Bayraktar SÇ (2018) Ubiquitous exposure to microfiber pollution in the air. Eur Phys J Plus 133:488. http://dx.doi.org/10.1140/epjp/i2018-12372-7
17. Dris R, Gasperi J, Mirande C, Mandin C, Guerrouachec M, Langlois V, Tassin B (2016) A first overview of textile fibers, including microplastics, in indoor and outdoor environments. Environ Pollu 1–6. http://dx.doi.org/10.1016/j.envpol.2016.12.013
18. Jönsson C, Arturin OL, Hanning AC, Landin R, Holmström E, Roos S (2018) Microplastics shedding from textiles—developing analytical method for measurement of shed material representing release during domestic washing. Sustainability 10:2457. http://dx.doi.org/10.3390/su10072457
19. Hearle JWS, Lomas B, Cookie WD (1998) Atlas of fiber Fracture and damage to textiles. Woodhead Publishing Ltd, London
20. Zhou H, Zhou L, Ma K (2020) Microfiber from textile dyeing and printing wastewater of a typical industrial park in China: Occurrence, removal and release. Sci Total Environ 739. https://doi.org/10.1016/j.scitotenv.2020.140329
21. Almroth BMC, Åström L, Roslund S, Petersson H, Johansson M, Persson NK (2018) Quantifying shedding of synthetic fibers from textiles; a source of microplastics released into the environment. Environ Sci Pollu Res 25:1191–1199. https://doi.org/10.1007/s11356-017-0528-7
22. Hernandez E, Nowack B, Mitrano DM (2017) Synthetic textiles as a source of microplastics from households: a mechanistic study to understand microfiber release during washing. Environ Sci Technol 51(12):7036–7046. https://doi.org/10.1021/acs.est.7b01750

23. Zambrano MC, Pawlak JJ, Daystar J, Ankeny M, Cheng JJ, Venditti RA (2019) Microfibers generated from the laundering of cotton, rayon and polyester based fabrics and their aquatic biodegradation. Marine Pollu Bull 142:394–407. http://dx.doi.org/10.1016/j.marpolbul.2019.02.062
24. Cooke WD, Arthur DF (1981) 10—a simulation model of the pilling process. J Text Inst 72(3):111–120. https://doi.org/10.1080/00405008108631637
25. Okubayashi S, Bechtold T (2005) A pilling mechanism of man-made cellulosic fabrics—effects of fibrillation. Textile Res J 75(4):288–292. https://doi.org/10.1177/0040517505054842
26. Cesa FS, Turra A, Checon HH, Leonardi B, Baruque-Ramos J (2019) Laundering and textile parameters influence fibers release in household washings. Environ Pollu 257:113553. https://doi.org/10.1016/j.envpol.2019.113553
27. Napper IE, Thompson RC (2016) Release of synthetic microplastic plastic fibres fromdomesticwashingmachines: Effects of fabric type and washing conditions. Mar Pollut Bull 112(1–2):39–45. https://doi.org/10.1016/j.marpolbul.2016.09.025
28. O'Loughlin C (2018) Fashion and Microplastic Pollution, Investigating microplastics from laundry. Ocean Remedy. https://cdn.shopify.com/s/files/1/0017/1412/6966/files/Fashion_and_Microplastics_Ocean_Remedy_2018.pdf. Accessed 15 April 2020
29. Yang L, Qiao F, Lei K, Huiqin Li Y, Kang SC, An L (2019) Microfiber release from different fabrics during washing. Environ Pollut 249:136–143. https://doi.org/10.1016/j.envpol.2019.03.011
30. Microfiber shedding−Topic FAQ, Outdoor Industry Association, European Outdoor group. https://static1.squarespace.com/static/5aaba1998f513028aeec604c/t/5db83b049c1c7a6e4cd66432/1572354825000/Microfiber+Shedding+FAQ+FINAL.pdf. Accessed 17 Aug 2020
31. Saxena S, Raja ASM, Arputharaj A (2017) Challenges in sustainable wet processing of textiles", s.s. muthu (ed.), textiles and clothing sustainability, textile science and clothing technology, springer science + business media, Singapore. http://dx.doi.org/10.1007/978-981-10-2185-5_2
32. Xia Xu, Hou Q, Yingang Xue YJ, Wang LP (2018) Pollution characteristics and fate of microfibers in the wastewater from textile dyeing wastewater treatment plant. Water Sci Technol. https://doi.org/10.2166/wst.2018.476
33. Bishop DP (1995) Physical and chemical effects of domestic laundering processes. In: Carr CM (eds) Chemistry of the textiles industry. Dordrecht: Springer. http://dx.doi.org/10.1007/978-94-011-0595-8_4
34. Galvão A, Aleixo M, De Pablo H, Lopes C, Raimundo J (2020) Microplastics in wastewater: microfiber emissions from commonhousehold laundry. Environ Sci Pollut Res. https://doi.org/10.1007/s11356-020-08765-6
35. Kelly M, Lant NJ, Kurr M, Grant Burgess J (2019) Importance of water-volume on the release of microplastic fibers from laundry. Environ Sci Technol 53:11735–11744. https://doi.org/10.1021/acs.est.9b03022
36. Laitala K, Klepp IG, Henry B (2018) Does use matter? comparison of environmentalimpacts of clothing based on fiber type. Sustainability 10:2524. http://dx.doi.org/10.3390/su10072524
37. Henry B, Laitala K, Klepp IG (2019) Microfibres from apparel and home textiles: Prospects for includingmicroplastics in environmental sustainability assessment. Sci Total Environ 652:483–494. https://doi.org/10.1016/j.scitotenv.2018.10.166
38. Preferred Fiber and Material, Market report (2019) Textile Exchange. https://store.textileexchange.org/product/2019-preferred-fiber-materials-report/
39. Erkoç M, Özkarademir G, Babaarslan O (2019) Microfiber waste load analysis in denim fabric laundering process. https://www.researchgate.net/publication/332697582
40. Lant NJ, Hayward AS, Peththawadu MMD, Sheridan KJ, Dean JR (2020) Microfiber release from real soiled consumer laundry and the impact of fabric care products and washing conditions. PLOS One 15(6):e0233332. https://doi.org/10.1371/journal.pone.0233332
41. Özkan İ, Gündoğdu S (2020) Investigation on the microfiber releaseunder controlled washings from the knitted fabrics produced by recycled and virgin polyester yarns. J Textile Inst. https://doi.org/10.1080/00405000.2020.1741760

42. Frost H, Zambrano MC, Leonas K, Pawlak JJ, Venditti RA (2020) Do recycled cotton or polyester fibers influence theshedding propensity of fabrics during laundering?. AATCC J Res 7(1):32–41. https://doi.org/10.14504/ajr.7.S1.4
43. Roos S, Levenstam O, Hanning AAC (2017) Microplasticsshedding frompolyester fabrics. Mistra Future Fash Rep Num 2017:1
44. De Falco F, Di Pace E, Cocca M, Avella M (2019) The contribution of washing processes of synthetic clothes to microplastic pollution. Scient Rep 9:6633. https://doi.org/10.1038/s41598-019-43023-x
45. De Falco F, Pia Gullo M, Gentile G, Di Pace E, Cocca M, Gelabert L, Brouta-Agnesa M, Rovira A, Escudero R, Villalba R, Mossotti R, Montarsolo A, Gavignano S, Tonin C, Avella M (2018) Evaluation of microplastic release caused by textile washing processes of synthetic fabrics. Environ Poll 236:619–925. https://doi.org/10.1016/j.envpol.2017.10.057
46. Spencer David J (2001) Knitting technology-a comprehensive handbook and practical guide, 3rd Edn. Woodhead Publishing Ltd, Cambridge, England
47. Cai Y, Mitrano DM, Heuberger M, Hufenus R, Nowack B (2020) The origin of microplastic fiber in polyester textiles: the textile production process matters. J Clean Prod. https://doi.org/10.1016/j.jclepro.2020.121970
48. Lord PR (2003) Handbook of yarn production Technology, science and economics. Woodhead Publishing Ltd, Cambridge
49. Doustaneh AH, Mahmoudian SH, Mohammadian M, Jahangir A (2013) The effects of weave structure and yarn fiber specification on pilling of woven fabrics. World Appl Sci J 24(4):503–506. https://doi.org/10.5829/idosi.wasj.2013.24.04.186
50. Lewin M (2007) Handbook of fiber chemistry: third edition. Taylor and Francis Group, London
51. Abdullah I, Blackburn RS, Russell SJ, Taylor J (2006) Abrasion phenomena in twill tencel fabric. J Appl Polym Sci 102:1391–1398. https://doi.org/10.1002/app.24195
52. Fatma K, Önder E, Özipek B (2003) Influence of varying structural parameters on abrasion characteristics of 50/50 wool/polyester blended fabrics. Textile Res J 73(11):980–984
53. Anderson CA, Leeder JD, Taylor DS (1972) The role of torsional forces in the morphological breakdown of wool fibres during abrasion. Wear 21
54. Study on the influence of fibres and fabrics properties in relation with fibres loss Report on the influence of commercial textile finishing, fabrics geometry and washing conditions on microplasticsrelease. MERMAIDS
55. Cai Y, Yang T, Mitrano DM, Heuberger M, Hufenus R, Nowack B (2020) Systematic study of microplastic fiber release from 12 different polyester textiles during washing. Environ Sci Technol 54(8):4847–4855. https://doi.org/10.1021/acs.est.9b07395
56. De Falco F, Gennaroa G, Robertoa A, Emanuelaa M, Emiliaa DP, Veronicaa A, Maurizioa A, Mariacristinaa C (2018) Pectin based finishing to mitigate the impact of microplastics released by polyamide fabrics. Carbohydr Polym 198:175–180. https://doi.org/10.1016/j.carbpol.2018.06.062
57. De Falco F, Cocca M, Guarino V, Gentile G, Ambrogi V, Ambrosio L, Avella M (2019) Novel finishing treatments of polyamide fabrics by electrofluidodynamic process to reduce microplastic release during washings. Polym Degrad Stab 165:110–116
58. Bresee RR (1986) General effects of ageing on textiles. J Am Inst Conser 25(1):39. https://doi.org/10.2307/3179413
59. Hartline N, Bruce N, Karba S, Ruff E, Sonar S, Holden P (2016) Microfiber masses recovered from conventional machine washing of new or aged garments. Environ Sci Technol 50(21):11532–11538. https://doi.org/10.1021/acs.est.6b03045
60. Bruce C, Hartline N, Karba S, Ruff E, Sonar S (2018) Microfiber pollution and the apparel industry. University of California, Santa Barbara, Bren School of Environmental Science & Management
61. Lamichhane G (2018) Analysis of microfibers in waste water from washing machines. Metropolia University of Applied Sciences, Bachelor of Engineering, Environmental Engineering Thesis. https://www.theseus.fi/bitstream/handle/10024/141278/Lamichhane_Ganesh.pdf?sequence=1&isAllowed=y. Accessed 15 April 2020

Current State of Microplastics Research in SAARC Countries—A Review

K. Amrutha, Vishnu Unnikrishnan, Sachin Shajikumar, and Anish Kumar Warrier

Abstract In recent decades, plastic has become an inevitable part of human life. It is found to be more suitable for various applications than any other materials due to its longevity, being lightweight, and versatility. However, it comes with its own set of problems to the nature which is mainly due to its non-biodegradability. Microplastics are small-sized (<5 mm) debris of plastics. They are ubiquitous in our environment and are known to exist in all the five spheres of our planet. There is a growing concern among the microplastic researchers about the fate of these tiny particles of plastics which are floating in the oceans and suspended in the atmosphere. Since the last two decades, significant amount of work has been reported on microplastic pollution and its effects on the biota in different parts of the world. In this chapter, we present a detailed review of microplastic studies carried out by SAARC nations (Afghanistan, Bangladesh, Bhutan, India, the Maldives, Nepal, Pakistan, and Sri Lanka). Sixty research papers were reviewed and it was found that India had carried out significant work in this research area. Pakistan, Bangladesh, Maldives, Sri Lanka, and Nepal have reported few studies. Other SAARC nations such as Bhutan and Afghanistan have yet to report the presence and effects of microplastics in their environment. There are no significant studies made on the presence and distribution of airborne microplastics and its impact on human health, influence of microplastics in agricultural and forest soils and how they affect the soil properties. Besides, there are no studies to check the efficiency of the wastewater treatment plants in major and minor cities of most of the SAARC nations. Studies have shown that the SAARC nations account for nearly a quarter of the world's population contribute a significant amount of plastic waste to the global plastic pollution. Therefore, a thorough understanding of microplastic pollution in these countries is essential that can be utilized for planning and implementing suitable legislative measures to safeguard the environment.

K. Amrutha · V. Unnikrishnan · S. Shajikumar · A. K. Warrier (✉)
Manipal Institute of Technology, Manipal Academy of Higher Education, Manipal 576104, India
e-mail: anish.warrier@manipal.edu

A. K. Warrier
Centre for Climate Studies, Manipal Academy of Higher Education, Manipal 576104, India

S. S. Muthu (ed.), *Microplastic Pollution*, Sustainable Textiles: Production, Processing, Manufacturing & Chemistry, https://doi.org/10.1007/978-981-16-0297-9_2

Keywords Microplastics · Pollution · Environment · Biota · Aquatic · Toxicity · Ecosystem · SAARC

List of Abbreviations

PP	Polypropylene
PA	Polyamide
PET	Polyethylene terephthalate
PS	Polystyrene
PE	Polyethylene
PVC	Polyvinylchloride
PU	Polyurethane
PAK	Polyacrylate
PVS	Polyvinyl stearate
EPC	Ethylene/propylene copolymer
SBR	Styrene/butadiene rubber
EPDM	Ethylene/propylene/diene rubber
AR	Alkyd resin
HDPE	High-density polyethylene
LDPE	Low-density polyethylene
NY	Nylon
PES	Polyester
PVCA	Vinyl chloride/Vinyl acetate Copolymer
ABS	Acrylonitrile/butadiene/styrene Copolymer
PVK	Poly (N-vinyl carbazole)
PEVA	PVC/Acrylic alloy, and Poly(ethylene-co-vinyl acetate)
PVF	Polyvinyl formal
EE	Epoxy epichlorhydrin
PP-PE	Poly(ethylene:propylene:diene) Copolymer
PVA	Polyvinyl acetate
CL	Cellulose
CP	Cellophane
PAE	Polyarylether
RY	Rayon
PB	Polybutadiene
PVB	Polyvinyl benzoate
TPA	Terephthalic Acid
BPA	Bisphenol A
PC	Polycarbonate

1 Introduction

It has been increasingly difficult for the present-day generation to imagine a world without plastics. Plastic materials are durable and non-biodegradable and these properties make it more popular to the mankind and at the same time, a villain to the ecosystem. The longer residence time of plastic materials in the natural environment due to these properties poses a serious threat to the ecology [131, 134]. As time passes, various factors, either biotic or abiotic, begin to degrade these materials to smaller particles [2] and the extent of degradation depends on their physicochemical properties [22]. Plastic is a common word applied to a group of polymers out of which significant classes are polyethylene, polystyrene, polyvinyl chloride, polyethylene terephthalate, polyurethane and polypropylene [63]. In recent decades, the generation of plastic waste has escalated at a shocking rate globally. Out of the 275 million metric tons of plastic produced in 192 coastal countries in 2010, approximately 4.8–12.7 million metric tonnes have reached the ocean [50]. The authors emphasized the importance of proper waste management systems and stressed on the serious consequences if the present scenario continues.

In modern times, environmental scientists are deeply concerned about 'microplastics' (Mps). They are small materials of plastic with a diameter <5 mm. Depending on their mode of formation, they are classified as primary and secondary microplastics [35, 69]. Primary microplastics are tiny plastics particles, produced for a particular function and used as such, that are released into the environment [15, 35]. One of the sources of these particles is facial cleansers, which contain polyethylene and they cannot be trapped by wastewater treatment plants resulting in them ending up into the oceans [33]. Some other sources of primary microplastics are manufactured pellets used in plastic production or the feedstock, industrial abrasives for sandblasting, plastic powders used in moulding, micro-beads used in the formulation of cosmetics [35, 78, 142]. Smaller plastic materials resulting from the degradation of large-sized plastics by the action of winds, UV radiation, currents, and microbes are secondary microplastics [35, 69]. Microplastics can be categorized, based on their morphology, into films, fragments, fibres, foams, and pellets [76, 64, 99]. In the past decade, the presence of microplastics have been documented from different environments: estuarine [97], riverine [116], lake [72], beach [42], polar [125], mangroves [26], road dust [43], soils [136], and even the atmosphere [141].

Due to its size, the harmful impacts of microplastics are much more significant than that of large-size plastic waste. Several negative effects on biota due to microplastics were reported including inhibited photosynthesis [13], weight-loss due to starvation [11], reduced filtration [133], impact on feeding and digestion [20, 35], and even mortality [71]. The ingestion of these particles mistaken as food by the aquatic organisms at the lower levels of food chain and their subsequent transfer through the food chain is a serious ecological issue [135]. As microplastics enter the food chain, the health of humans [124] and predators who feed on these organisms is under threat [41]. This is supported by the fact that microplastics have been isolated from sea birds [18], benthic invertebrates [81], fish guts [60], harbour seals—a marine mammal [95], etc. They are reported to occur in oceans worldwide [8, 27, 58, 100] with the gyres being one of the prominent centres for microplastics accumulation [135]. Microplastics can also be a vector for other environmental contaminants such as antibiotics (González-Pleiter et al. [38] in press), persistent organic pollutants [119], metals, invasive species, and associated bacterial pathogens [82]. This severely affects the aquatic organisms which ingest the microplastics [11]. Besides, the microorganisms can develop biofilms and colonies on these microplastics due to the presence of a hydrophobic surface [139]. Another essential factor to consider is the rate of degradation of Mps. They can be further degraded by biotic or abiotic factors [64] among which the abiotic factor, such as photo-oxidation, is the primary mechanism for most of the abundant polymer types [44]. Fragmentation of microplastic further reduces their size and increases their surface area, which makes them more susceptible to further degradation into smaller particles [44]. The microplastic study has high relevance as the United Nations has announced a programme 'Transforming our World: the 2030 Agenda for Sustainable Development'.

In 1985, the South Asian Association for Regional Cooperation (SAARC) was established. The SAARC comprises of eight member countries: Afghanistan, Bangladesh, Bhutan, India, Maldives, Nepal, Pakistan, and Sri Lanka. These countries are also known as the developing nations in terms of their economic activities. As these activities have increased in recent decades, the level of plastic pollution has also risen. For example, during 2018, the SAARC nations alone produced 17–20 million tonnes (MT) of plastics. Since 2016, these countries have dumped nearly 26.72 MT of plastic wastes. India is one of the greatest economies and highly populated country in SAARC and generates ~26,000 tonnes of plastic/day and around 80% of the total plastic produced in the country ends up in the garbage [59]. The World Bank has predicted that the plastic production in India will reach 20 MT by 2020, and the amount of plastic waste thus generated will be high. Table 1 shows the details of the generation of plastic waste in some of the SAARC nations since 2010 [50]. Pakistan (6.41 MT) leads the table followed by India (4.49 MT's) and the least was recorded in Maldives (43,134 tonnes) as per the 2010 data. Bangladesh and Sri Lanka accounted for 1.89 MT's and 2.62 MT's of plastic waste.

In this chapter, we present a detailed review of the present knowledge about the occurrence, distribution, and fate of microplastics in different environmental compartments of the SAARC countries. This exercise aimed to find the status of microplastics research in these countries and to find knowledge gaps that can be

Table 1 Quantum of plastic waste generated by a few SAARC nations in 2010 [50]

Country	Plastic waste generation
Pakistan	6.41 MT
India	4.49 MT
Sri Lanka	2.62 MT
Bangladesh	1.89 MT
Maldives	43,134 T

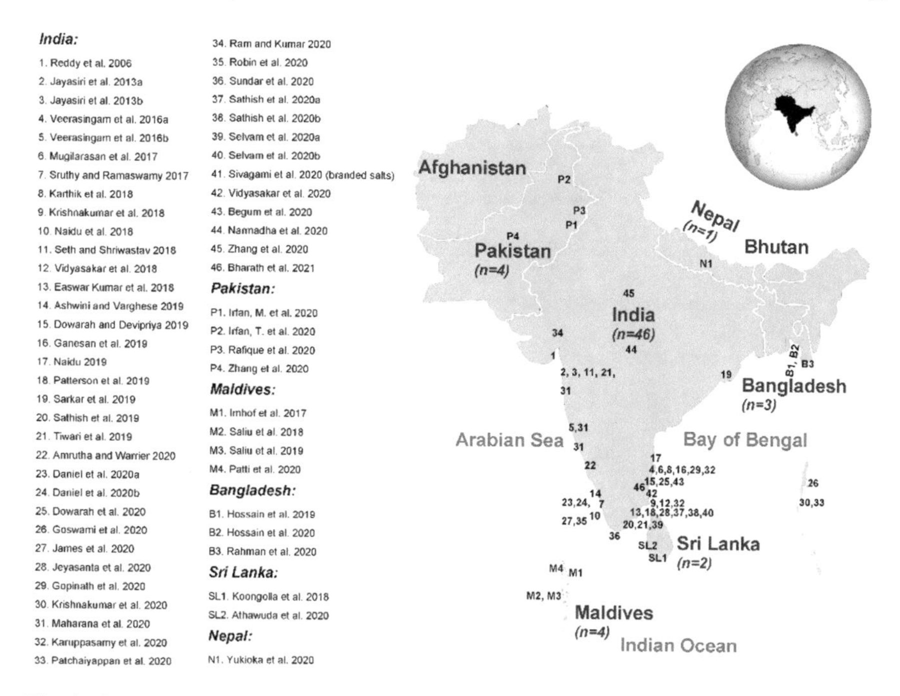

Fig. 1 General map of the SAARC nations showing the locations of all microplastic studies reviewed in this chapter

considered for future studies in this field. For this purpose, we searched for literature published till 30 September 2020. The search was made using online platforms such as Google Scholar, Web of Science, ScienceDirect, and Scopus. The following keywords and their combinations were used to search for literature: microplastics, environmental pollution, FTIR, biota, sediments, water). Besides, the names of the SAARC nations were tagged along with the keywords as mentioned earlier. A total of 60 publications from the SAARC nations were found in which microplastics research work was documented from terrestrial and marine ecosystems (Fig. 1).

2 The Knowledge of Microplastics Research in SAARC Nations

Compared to the amount of work that has been carried on different facets of microplastic research in other parts of the world, works conducted in the SAARC nations are relatively few. However, there is a rising trend seen in the number of data

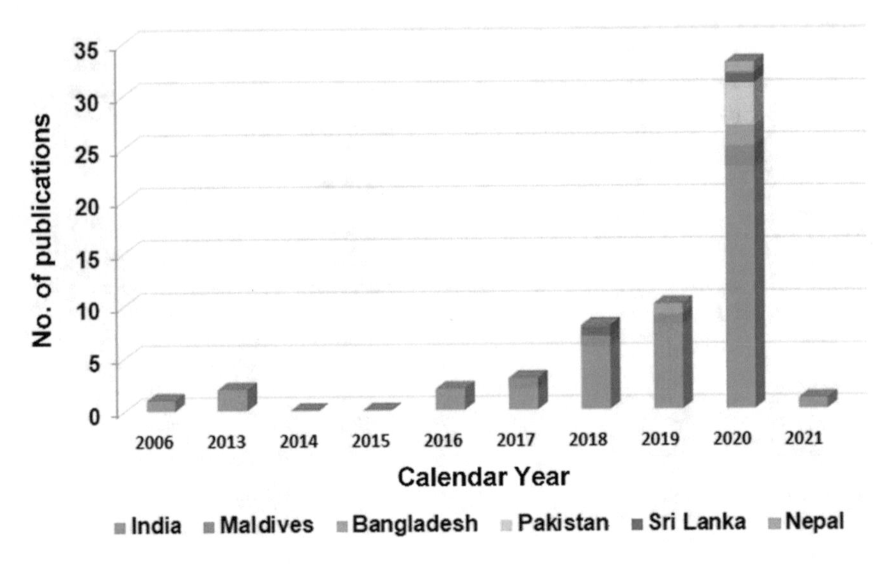

Fig. 2 Total number of publications pertaining to microplastics published by researchers from the SAARC nations. A significant rise is seen between 2018 and 2020 with India leading the research on terrestrial and aquatic ecosystems

that is generated in the last two years justified by the number of peer-reviewed publications (Fig. 2, Tables 2 and 3). In the following section, we discuss the occurrence, distribution, and the ultimate fate of microplastics in the different environments (Fig. 3) of the SAARC nations.

3 India

3.1 Coastal Environment

The global plastic production is exponentially increasing on an annual basis. In 2015, the global plastic production exceeded 320 million tonnes [101]. These materials will persist in nature for a very long period and will ultimately reach the oceans through various pathways including different water bodies, surface runoff, wind transportation, and atmospheric deposition [74]. A study conducted on plastic pollution in a water channel in the South Juhu creek, Mumbai, suggested that a significant amount of plastic wastes is being carried from terrestrial environments to the seas/oceans through connected water channels [75]. Plastics from these sources undergo various processes, including beaching, drifting, and settling before they reach temporary reservoirs on coastal environments [140]. Coastlines act as a significant temporary/permanent reservoir for the accumulation of plastics released to sea/oceans

Table 2 A summary of the research work carried out on microplastics in different environmental settings of India

Sl. no	Location	Type of sample	Environmental setting	Methodology adopted	Size range (mm)	Average microplastics concentration	Polymer types	Reporting units	References
1	Gujarat	Sediments	Marine	Olympus microscope, FTIR, Scanning electron microscope	Small plastic fragments	81 mg kg^{-1}	PU, NY, PS, Glass wool	mg kg^{-1}	[96]
2	Mumbai	Sediments	Beach	Handpicked	<5 mm, 6–20 mm, 21–100 mm	11.67 ± 8.83 items/m^2 by items 3.24 ± 0.92 g/m^2 by weight	–	Items/m^2 g/m^2	[54]
3	Mumbai	Sediments	Beach	Handpicked	1–5 mm	–	–	–	[55]
4	Chennai	Sediments	Coastal	NIKON Stereoscopic microscope SMZ1500 coupled with a digital camera; ATR-FTIR	2–5 mm	1200 pieces	PE, PP	No. of pieces	[127]
5	Goa	Sediments	Beach	Stereoscope microscope, ATR-FTIR	<5 mm	3000 pieces	PE, PP	No. of pieces	[128]
6	Chennai and Thinnakkara island	Sediments	Beach	NIKON Stereoscopic microscope	<5 mm	Chennai: 201 Thinnakkara: 603	–	No. of particles	[79]
7	Kerala	Sediments	Lake	Compound microscope, microRaman spectroscopy	<5 mm	252.8 particles/m2	HDPE, LDPE PP, PS	Particles/m^2	[118]
8	Southeast coast of India	Sediments Fish	Beach Marine	Stereomicroscope, ATR-FTIR	Mps: 0.3–4.75 mm Mesoplasti-cs: 4.75–9.5 mm	46.6 ± 37.2/m 210.1% of the 79 fishes representing 5 species.	PE, PP, PS, PE,	Particles/m^2	[60]
9	Gulf of Mannar	Marine debris	Marine	Leica binocular stereozoom microscope, ATR-FTIR	–	less dominated by microplastics	PE, PVC, PS, Nylon	–	[67]
10	Kerala	Benthic invertebrates from coastal waters	Coastal	DXR Raman Microscope. Epifluorescent Microscopy	–	–	PS	–	[81]

(continued)

Table 2 (continued)

Sl. no	Location	Type of sample	Environmental setting	Methodology adopted	Size range (mm)	Average microplastics concentration	Polymer types	Reporting units	References
11	India	Salt	Coastal	Digital Microscope, CMOS sensor digital camera; µ-FTIR, FTIR Microscope	<5 mm	103–56 particles/kg	Polyester PE, PET, PS, PA	Particles/kg	[112]
12	Rameswaram Coral Island	Sediments	Coastal	Stereo zoom binocular Microscope; ATR-FTIR	Micro, Meso, Macro	403 pieces	PP, PE, PS, NY, PVC	No. of pieces	[130]
13	Tuticorin	Fish	Marine	Stereo Microscope (Coslab ZSM-111) FTIR analysis	–	12 out of 40 fishes	PE, PP	–	Easwar [68]
14	Kerala	Sediments	Beach	FTIR-ATR (Perkin Elmer-Spectrum 2 with Spectrum 10 software); SEM	1–5 mm	70.15 items/kg (July 2017) 120.85 items/kg (May 2018)	PE, PP, PS, PCU	Items/kg	[3]
15	Puducherry	sediments	Beach	Zeiss Primo Star microscope (40X); Raman spectroscope	1–5 mm	72.03 ± 19.16 items/100 g	PP, PE, PET, Others	Items/100 g	[29]
16	Chennai	Surface waters Packed -drinking water Groundwater	Marine, Lake	Optical microscope 40 × with digital camera SEM-EDX; FTIR-ATR	<5 mm	66 particles	PET, PA	No: of particles	[36]
17	Chennai	Green Mussels	Marine	DXR Raman microscope (thermo scientific, USA; 40–80x), DXR Raman spectroscopy	–	0.9 ± 0.3 items/10 g to 3.2 ± 3.2 items/10 g	PS	–	[80]
18	Tuticorin	Oysters	Marine	Stereomicroscope, FTIR-ATR, SEM-EDX	0.005–5 mm	0.81 ± 0.45 items/g	PE, PP, PA, PES, Others	Items/g	[89]
19	Ganges River	Sediments	Riverine	Microscope (Nikon Eclipse Ci fitted with camera Nikon DS-Fi2), ATR-FTIR	<5 mm	107.57 to 409.86 items/kg	PET, PE, PP, PS	Items/kg	[105]

(continued)

Table 2 (continued)

Sl. no	Location	Type of sample	Environmental setting	Methodology adopted	Size range (mm)	Average microplastics concentration	Polymer types	Reporting units	References
20	Tamil Nadu	Sediments	Beach	FTIR-ATR, Dissecting microscope (40X); SEM-EDX	0.5–3 mm	309 ± 184 to 76 ± 72 items/kg	PE, PP, NY, PES, PS	Items/kg	[106]
21	Mumbai Tuticorin Dhanushkodi	Sediments	Beach	FTIR-ATR, SEM-EDS, Fluorescence microscopy	0.036–5 mm	Mumbai: 220 ± 50 MP kg^{-1} Tuticorin: 181 ± 60 MP kg^{-1} Dhanushkodi: 45 ± 12 MP kg^{-1}	PE, PET, PS, PP, PVC, Others	kg^{-1}	[126]
22	Karnataka	Surface waters Sediments Soil	River	Nikon Stereozoom Microscope (40 × magnification); FTIR-ATR	1-5 mm & 0.3-1 mm	Surface waters: 288 pieces/m^3. Sediments: 96 pieces/kg	PE, PET, PP, PVC	Pieces/m^3 Pieces/kg	[1]
23	Coastal waters of Cochin, India	White Shrimps	Coastal	Stereo Microscope; FTIR-ATR	157–2785 μm	0.39 ± 0.6 microplastics/shri-mps	PET, PA, PE, PP	Microplastics/shrimps	[23]
24	Kerala	Fish	Marine	Stereo microscope, ATR-FTIR	100–5000 μm	Edible tissues: 0.005 ± 0.02 items/g Inedible tissues: 0.054 ± 0.098 items/g	PP, PE, EPDM PS, PU	Items/g	[24]
25	Pondicherry	Bivalves	Estuarine	Nile Red Staining, Fluorescent Microscope, (Olympus CX41), Raman Spectroscopy	<100 μm	0.18 ± 0.04 to 1.84 ± 0.61 items/g 0.50 ± 0.11 to 4.8 ± 1.39 items/individual	PU, PVCA, PVC, PES, PET, ABS, SBR, PVK, PEVA	Items/g Items/individual	[30]

(continued)

Table 2 (continued)

Sl. no	Location	Type of sample	Environmental setting	Methodology adopted	Size range (mm)	Average microplastics concentration	Polymer types	Reporting units	References
26	Port Blair Bay	Water Sediment Zooplankton Fish	Marine	Stereomicroscope, ATR-FTIR	<5 mm	Water: 0.93 ± 0.59 particles/m^3 Sediments: 45.17 ± 25.23 particles/kg Zooplankton: 0.12 ± 0.07 pieces/zooplankter Fish: 10.65 ± 7.83 particles/specimen	Ionomer-surlyn Polyeteramide Acrylic Polyphenylene-sulphide EVOH Acrylonitrile NY Ethylene vinyl- acetate Polyisoprene PU, PVC	Particles/m^3 Particles/kg Pieces/zooplankton Particles/specimen	[40]
27	Cochin	Surface water Sediments Fish	Marine	Stereomicroscope (40x), Raman Spectrometer	<1 mm 1–5 mm.	30 MPs from fishes	PE, PP	–	[51]
28	Tuticorin	sediments	Beach	Dissecting microscope (40X), FTIR-ATR	0.05–5 mm	25 ± 1.58 to 83 ± 49 items/m^2	PE, NY, PET, PS, PP, PVC	Items/m^2	[56]
29	Chennai	Sediments water samples	Lake	Stereomicroscope, ATR-FTIR, SEM- EDX	<5 mm	5.9 particles/L in water 27 particles/kg in sediments	PP, PS	Particles/L Particles/kg	[39]
30	Andaman and Nicobar	Sediments	Beach	Stereozoom Microscope with online digital camera, ATR-FTIR	>0.1–0.45 μm micro	73 to 151 items	PVC, PS, Nylon, PE, PP	No: of particles	[67]
31	Maharashtra, Karnataka and Goa	Sediments	Beach	Stereo zoom microscope, ATR-FTIR	1–5 mm	43.6 ± 1.1 to 346 ± 2 items/m^2	PE, PP	Items/m^2	[73]
32	Tamilnadu	Fish	Marine	Stereomicroscope (10 × & 20x), FTIR-ATR, SEM	1.4–9.3 mm	20 particles out of 190 fishes. Avg. ingestion: 8.95%	PE, PET, PA	No. of particles Percentage	[61].

(continued)

Table 2 (continued)

Sl. no	Location	Type of sample	Environmental setting	Methodology adopted	Size range (mm)	Average microplastics concentration	Polymer types	Reporting units	References
33	South Andaman Island	Sediments	Beach	Nile Red Staining, Fluorescent microscopy, Raman Spectrometer	>100–1000 µm	414.35 ± 87.4 particles/kg	PP, PVC, PVF PB, Polysulfide Melamine PVB, Nylon-6 EE, ABS	Particles/kg	[88]
34	Ahmedabad	Sediments	River	Microscope(50x), SEM	75–212 µm 212 µm to 4 mm	47.1 mg 4 mg	–	mg	[94]
35	Southwest coast of India	Sediments water Fish	Coastal and Marine	Stereomicroscope fitted with a digital camera, FTIR-ATR.	0.3–4.75 mm	40.7 ± 33.2 particles/m^2 (22 beaches) Coastal waters: 1.25 ± 0.88 particles/m^3 Microplastics in 15 fishes out of 70 fish samples	PE, PP, PA, PS, Others	Particle/m^3	[98]
36	Kanyakumari	Sediments	Coastal	Binocular microscope (40x)	<5 mm	343 particles/50 g dry sediment	–	Particles/50 g dry sediment	[123]
37	Tuticorin	Fish Water	Marine	Dissecting Microscope (40x), FTIR-ATR, SEM-EDX	<5 mm	3.1–23.7 items/L in water 0.11–3.64 items/individual	PE, PES, PA, PS, PP, PP-PE, PVA, Acrylic	Items/L Items/individual	[108]
38	Tuticorin	Salt	Coastal	Motic Stereoscopic microscope (40x), FTIR-ATR, SEM-EDX	<100 µm- >1000 µm	35–72 items/kg in sea salt 2–29 items/kg in bore-well salt	PE, PP, PES, PA	Items/kg	[107]
39	Tamil Nadu	Ground water Surface water	Coastal	stereoscopic microscope (XRZ3, Olympus; 20-80x), µ-FT-IR, Atomic Force Microscopy, ATR-FTIR	<5 mm	Groundwater: 4.2 particles/L Surface water: 7.8 particles/L	PA, PES, PP, PE, PVC, CL	Particles/L	[110]

(continued)

Table 2 (continued)

Sl. no	Location	Type of sample	Environmental setting	Methodology adopted	Size range (mm)	Average microplastics concentration	Polymer types	Reporting units	References
40	Tuticorin	Salt	Marine	Binocular stereo zoom microscope, μ-FTIR and Atomic Force Microscopy	<2 mm	–	NY, CL, PP, PE	–	[111]
41	India	Commercial sea Salt	Marine	Light microscope, Fluorescence Microscope, FTIR	5.2–3.8 μm	<700MPs/kg	CP, PR, PA, PAE, PE, PET	Mps/kg	[115]
42	Tamilnadu	Sediments	Beach	Stereozoom Microscope, FTIR-ATR	>2 mm	204 particles/kg	PE, NY, PVC	Particles/kg	[129]
43	Tamilnadu	Sediments	Estuary	Filtering, Fluorescence Microscopy	–	30.2, 9.4, 11.7 mg/kg	–	mg/kg	[9]
44	Nagpur	Dust	Roadside	Stereomicroscope (10 × 40x), ATR-FTIR, SEM	PM_{10} $PM_{2.5}$	50 to 120 particles/day	LDPE, Rayon Rubber, PS Polyaniline, Polyolefin, PVC	Particles/day	[84]
45	Patna	Dust	Indoor dust	HPLC-MS/MS	<5 mm	55 to 6800 μg (PET) <0.11 to 530 μg (PC)	PET, PC	μg	[141]
46	Tamilnadu	Water sediments	Lake	ATR FTIR	0.3–2 mm	Water: 28 items/km^2 Sediments: 309 items/kg	PP, PS, Nylon PVC, PE	Items/km2 Items/kg	[12]

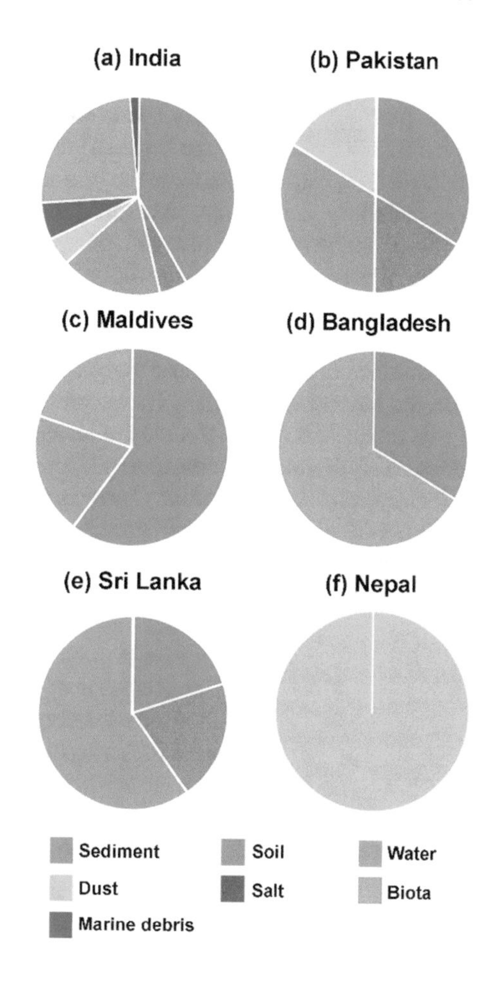

Fig. 3 Pie-chart distribution of microplastic research carried out on different environmental compartments by the SAARC nations

[140] and the coastal zone is considered as a significant microplastic hotspot as about half of the world's population resides in the vicinity of coasts [20]. Different works have suggested the negative of microplastics on the coastal ecosystem and the need to resolve the issues. India is surrounded by the Arabian Sea in the west, the Bay of Bengal in the east, and the Indian Ocean in the south. It has a coastline of nearly 7516.6 km (including islands) touching 13 States and Union Territories with a population density of 560 million [17]. The review by Saha et al. [102] discussed the problem of microplastic pollution along the Indian coast and expressed the need for more investigations on microplastic contamination in different coastal regions of India. On the western coast of India, microplastic investigations have been reported from the shores of Gujarat, Mumbai, Goa, Kerala, and Karnataka. Reddy et al. [96] wrote the pioneering work on plastic debris from Gujarat and Maharashtra coast. They studied the intertidal sediments of the world's largest shipping yard at Alangososiya, India, for the presence of small plastic debris. The obtained fragments

were the materials used by the construction industries and related activities and pinpointed the source of the pieces as ship breaking activities. Recreational beaches of Mumbai, Maharashtra have been studied for plastic pollution by Jayasiri et al. [54, 55]. They found that the beaches are predominated by plastic litter. Their work concluded terrestrial input to be the prominent source of these materials. The activities including recreational, tourism, religious, and fishing were the major sources of plastic pollution in the area [54, 55]. They also carried out the size fractionation studies of plastics obtained from the beach sediment samples. They found that almost 41.85% of the microplastic particles, that is, the lion shares of the obtained plastic debris in size range of 1–5 mm. The study conducted by Maharana et al. [73] in Maharashtra beaches also observed the same—higher abundance of microplastics than the large-sized plastics. The quantity of tiny-sized plastic pieces observed in the latter study is almost double the former, which may be due to the high level of industrialization and seasonal tourism. Another study observed the higher abundance of microplastics in the Mumbai Coast compared to two other beaches they studied in the Bay of Bengal coast [126]. Further, [53] concluded that the plastic pellets are a trap for various cyclodiene compounds, including DDT, PCB, and HCH. They also suggested that pellet study can be used to study the temporal variability for specific micropollutants (Table 2).

Marine litter and its pathways in three beaches in the Mangalore Coast, Karnataka has been investigated by Sulochanan et al. [122] and suggested the need for a proper plastic management system for the conservation of the coastal ecosystem. The study highlighted that the physical forces result in the fragmentation of the litter and its adverse effect on the food web. From the research on marine litter in Karnataka coast, they reflected the role of rivers in carrying plastics to the coast [122, 121]. Sridhar et al. [117, 73] have also reported plastic pollution in Karnataka. Microplastic studies along the Goa coast were conducted by Veerasingam et al. [127]. The authors studied the characteristics and surface degradation properties of microplastics pellets (MPPs). They observed striking seasonal variations of their distribution along the coast. According to them, the southwest monsoonal rainfall played a significant role in transporting the microplastic particles to the beaches. In 2020, Maharana et al. also observed the high abundance of Mps along the Goan coast. But according to their study, the Goan coast shows a lower level of plastic pollution compared to the other two coasts—Maharashtra and Karnataka coast—studied by them. Two studies from the Kerala Coast revealed the influence of fishing and recreational activities on the distribution of microplastics in the coastal environments/beaches [29, 56]. The study conducted by Jeyasanta et al. [56] observed a close relationship between mesoplastics and microplastics. Still, according to their research, there is no influence of the abundance of macroplastics on the concentration of small-sized plastic particles. They also stressed the urgency of a proper management system to control plastic waste, which may otherwise result in a high level of pollution on the beaches. Ashwini and Varghese [3] have applied an effective strategy, environmental forensic investigation, to identify the source of microplastics on the Nattika coast. The process consists of three levels matching the material identity with the possible material using tests; comparing the materials with similar macroplastics and finding the pathway

Table 3 A summary of the research work carried out on microplastics in different environmental settings of Pakistan, Bangladesh, Maldives, Sri Lanka, and Nepal

Sl. no	Location	Type of sample	Environmen-tal setting	Methodology adopted	Size range (mm)	Average microplastics concentration	Polymer types	Reporting units	References
M1	Maldives	Sediments	Marine	Stereomicroscope with digital camera, ATR-FTIR spectrometer	1–5 mm >5 mm	Long term accumulation: 1029 ± 1134 plastic particles/m^2 daily accumulation: 35.8 ± 42.5 plastic particles/m^2	PE, PP, PS, PVC, PU, Polyvinyl Alcohol	Plastic particles/m^2	[47]
SL1	Sri Lanka	Water beach sediments	Beach	ATR-FTIR, raman spectroscopy	1.5–2.5 mm and 3–4.5 mm	20 m distance: 42.1 items/m^2 10 m distance: 96.6 items/m^2	PE, PP, PS	Items/m^2	[65]
M2	Maldives	Sediments water	Marine	Stereomicroscope, ATR-FTIR, spotlight 200i FTIR microscopy system	>5 mm 1–5 mm	Water: 0.32 ± 0.15 particles/m^3 Sediments: 22.8 ± 10.5 particles/m^2	PE, PP, PS, PET, PA, PVC	Particles/m^3 Particles/m^2	[103]
B1	Northern Bay of Bengal	Fish	Marine	Microscope with digital camera, μ-FTIR	<500 μm 500 μm–1 mm 1–5 mm	3.20–8.72 item/species	PET, PA	Item/species	[46]
M3	Maldives	Coral reefs	Marine	Stereoscopic microscope, infrared Spectroscopy	<5 mm	Inside: 0.46 ± 0.15 items/m3 Outside: 0.12 ± 0.09 items/m3	PE, PP, PA, PS, PU	Items/m3	[104]

(continued)

Table 3 (continued)

Sl. no	Location	Type of sample	Environmen-tal setting	Methodology adopted	Size range (mm)	Average microplastics concentration	Polymer types	Reporting units	References
N1	Nepal	Road dust	Urban	Stereoscopic Microscope and digital camera, ATR-FTIR	100–5 mm	Kathmandu: 12.5 ± 10.1 pieces/m2	PE, PP, PS, PET, PAK, PVS, EPC, SBR, EPDM PU	Pieces/m2	[138]
B2	Northern Bay of Bengal	Shrimp	Marine	Microscope with digital camera, μ-FTIR	1–250 μm 250–500 μm 500 μm–1 mm 1–5 mm	P. monodon-3.40 ± 1.23 item/g M. monocerous-3.87 ± 1.05 item/g	PA rayon	Item/g	[46]
P1	Pakistan	Sediment water	Riverine	Stereomicroscope with digital camera, ATR-FTIR	300 μm-5 mm 150–300 μm	Water: 2074 ± 3651 MPs/m^3 Sediments: 3726 ± 9030 MPs/m^2	PE, PP, PS	MPs/m^3 MPs/m^2	[49]
P2	Pakistan	Sediment water	Lake	Light microscope (10x and × 40), FTIR	<5 mm	Water: 0.142 items/0.1 L Sediments: 1.04 items/0.01 kg	PE, PP, PET, PVC	Items/0.1 L Items/0.01 kg	[48]
M4	Maldives	Sediments	Marine	Stereozoom Microscope (30x) with Digital Microscope Imager	<5 mm	277.90 ± 24.98 MPs/kg	–	Mps/kg	[90]

(continued)

Table 3 (continued)

Sl. no	Location	Type of sample	Environmen-tal setting	Methodology adopted	Size range (mm)	Average microplastics concentration	Polymer types	Reporting units	References
B3	Bangladesh	Sediments	Beach	Microscope, ATR-FTIR	>300 μm	8.1 ± 2.9 Particles/kg	PP, PE, PS, PVC, PET, AR	Particles/kg	[93]
P3	Pakistan	Soil	Urban	Stereomicroscope, FTIR	300–5000 μm 50–150 μm 150–300 μm	4483 ± 2315 MPs/kg	PE, PET, PP	MPs/kg	[92]
P4	Pakistan	Dust	Indoor dust	HPLC-MS/MS	<5 mm	45–38000 μg (PET) <0.11–49 μg (PC)	PET, PC	μg	[141]
SL2	Sri Lanka	Water	Marine	Light microscope	0.30–1 mm 1–5 mm	139.77 ± 120.76 items/m3: size fraction by number 0.33 ± 0.13 mg/m^3: size fraction by weight	–	Items/m^3 mg/m^3	[4]

of the pollution. This study confirms that the primary source of microplastics in the beaches is the fragmentation of larger plastics that stay on the shore for an extended period before getting disturbed by wave/current actions. Their study also implied the role of rivers as essential pathways for transporting microplastics to the coast. A complete study was conducted by Robin et al. [98] on the coastal region of Kerala for understanding the presence of microplastics in the coastal waters and beach sediments. They observed a greater number of microplastic particles in the southern coast of the state. The presence of urban areas near to the sampling sites and river runoff was one of the key factors that affected microplastic distribution. Their analysis indicated the presence of harmful chemicals on the surface of microplastics. Further, microplastic distribution varied spatially, temporal and depth-wise in the water and sediment samples collected along a transect off Kochi [51]. The study demonstrated the positive correlation of monsoon on the abundance and distribution of microplastics.

On the eastern coast of India, most of the works are concentrated in the Tamil Nadu region. A study was conducted on microplastic pellets present in the surface sediments collected, before and after flooding, from the Chennai coast [128]. They investigated the surface features and age of these materials. They also studied the distribution, source, and polymer composition of these pellets. They observed that the abundance of pellets after flooding is more compared to pre-flooding, indicating the transportation pellets through rivers. A surprising fact is that plastic fragments were reported from an inland island with a low population effect—Nallathanni Island, Gulf of Mannar biosphere reserve of the southeastern coast of India [67]. This is due to the action of wind and waves. They observed that most of the marine debris had been dumped in the high tide region, which was attributed to the increased wave energy waves and landward flowing of wind pattern. The presence of plastic materials in the inner parts of the coral island observed during the study was due to the action of wind. The study observed a low concentration of Mps in the coastal sediments. The authors used the morphology of microplastics as a proxy for the identification of their origins. Another study by Vidyasakar et al. [130] also concluded the influence of aeolian transport on plastic distribution along the coast. The authors observed plastic pollution, both macro and microplastics in the sediments of Rameswaram Island, Gulf of Mannar. The study on marine litter on the beaches in the northern Gulf of Mannar region suggested the influence of population distribution, proximity to pilgrim centres, and southwest monsoon on microplastic abundance [37].

In another study, out of the 25 beach samples collected, a greater abundance of Mps is reported from a beach close to a river mouth [60]. They have observed a decrease in the number of microplastics on the beach as the distance from the nearest river mouth increases. Moreover, the authors also reported the presence of microplastic particles from a fish gut study. They concluded that the fish population in the region is severely affected by the presence of microplastics. A survey by Sathish et al. [106] on five beaches of Tamil Nadu found out that river input, religious and coastal activities, sewage inflow, etc., result in the accumulation of microplastics on the coast. Microplastic pollution in the Tuticorin beach and the Dhanushkodi beach of Tamil Nadu was mainly due to the industrial activities and fishing and tourist activities respectively [126]. Silver Beach, southern India showed slightly more significant plastic pollution than that on the Tuticorin coast [129]. The main reason that authors pointed out was the influence of the discharge of urban runoff into the Gadilam River and suggested that this river was the primary source of plastic debris on the beach. The study found out that 65% of the obtained plastic debris are mesoplastics, 18% micro, and 17% macro plastics materials. They also highlighted the influence of higher surface runoff during monsoon and aeolian deposits during the dry season in the distribution of plastics along the beach. Tourism and fishing activities are found to be the primary sources of plastic materials in the coastal region.

The presence of microplastic resin pellets from two beaches from the Tamil Nadu coast was documented by Mugilarasan et al. [79]. The Tinnakkara Island, even though it is remote, observed a higher abundance of pellets compared to the Chennai coast. The primary sources of these materials to the Tinnakkara Island were generated from external sources, that is activities away from the islands. The ocean currents and wind act as a carrier of these materials from the outside areas to the island. Because of the higher harbour and industrial activities, the microplastic resin pellets obtained from the Chennai coast are highly weathered and absorbed many environmental pollutants. Furthermore, secondary microplastics were obtained from the beach sediments of Kanyakumari, located in southernmost India [123]. Higher microplastics were obtained from tourist beaches, followed by harbours, coastal fishing villages/residential beaches, and undisturbed coastal areas, suggesting the influence of the different degrees of human activities on microplastic distribution.

Kaladharan et al. [57] studied the marine litter on different beaches along the coast of Peninsular India and the islands of Andaman and Nicobar and Lakshadweep. According to them, the garbage comprises mostly of plastics, and this problem can be reduced by social awareness. Plastic pollution has been reported from the coastal environment of the Andaman and Nicobar Islands even before [28], Das and Mohanty [25]. A higher concentration of plastic wastes was documented in the Nicobar Island sector and also the North and Middle Andaman Island sector [67]. According to the authors, currents play an important role in microplastic distribution in the coastal environment. The microplastics were also reported from the different stations along the South Andaman beach (mean $= 414.35 \pm 87.4$ particles per kilogram of beach sediments; [88]). The Port Blair Bay is also affected by microplastic pollution with an average abundance of 45.17 ± 25.23 particles per kilogram in sediments and 0.93 ± 0.59 particles per m^3 in surface seawater samples collected from the bay [40]. Begum et al. [9] investigated the sediments of three estuarine regions, namely, Uppanar, Vellar, and Coleroon, for microplastic pollution. They found a higher concentration of Mps in the sediments of the Tamil Nadu coastal zone [9]. Their results suggest that freshwater rivers are a potential source of Mps. Land-based sources including waste disposal from industries near the study area and recreational activities, are dominant sources of Mps in the study area compared to the sea-based sources.

Based on all the works, it is inferred that terrestrial sources of microplastics are more compared to marine sources. Fishing practices, including aquaculture and shipping activities, are the observed primary ocean-based sources of Mps. Tourist/recreational activities, beach littering, unplanned waste disposal, including the discharge of untreated sewage and industrial wastewater, and riverine input of plastic wastes can be the significant sources of land-based microplastics. Sharma and Chatterjee [113] did a review on global microplastic pollution in the marine environment in which they discussed the different aspects of microplastic pollution—their abundance, distribution, transfer, and effects. They emphasized the importance of regulating the use of plastics and the need for stringent rules and regulations in the production of plastics.

3.2 *Riverine Environment*

Most of the studies emphasize the importance of rivers as carrier systems of Mps to the ocean. Unfortunately, only a few studies have been reported on riverine microplastic pollution from India. Sarkaret al. [105] have studied the spatial distribution of mesoplastics and microplastics in the sediments along the lower reaches of the Ganges River. They also carried out the textural analysis, measured pH and specific conductivity, calculated organic carbon, total nitrogen, available nitrogen, and available phosphate. They observed a strong correlation between the abundance of microplastics and pollution traits. The river Ganga is heavily polluted with plastic debris, either in the form of films or fibres. The major reason attributed to this is the direct discharge of plastic waste into the river sewage. They emphasized the urgency of implementing measures to mitigate the pollution. As the river input of plastics is a crucial factor in microplastic pollution studies, the present scenario has to be changed. More researches on microplastic distribution along and across the whole river stretch should be conducted. Microplastic investigation in India should take a new dimension to attain a sustainable environment. Amrutha and Warrier [1] investigated the extent of microplastics contamination in the Netravathi River—an important west-flowing river draining the Western Ghats. The team has investigated the distribution of microplastics of the river from its source to its sink, and they have collected surface water, sediments, and soil samples. They observed a higher abundance of Mps in the water samples collected from the sampling station closer to the Arabian Sea. They reported a large amount of polyethylene terephthalate fibres in the samples collected from the river which are mainly derived from clothes. Discarding of cloth wastes and the removal of fibres from clothes during washing is a possible source of this. Ram and Kumar [94] have collected river samples from four stations in the Sabarmati River for the analysis of microplastics in it. A higher abundance of Mps was observed near a landfill site, and therefore they inferred that the major source of Mps in the river is illegal dumping. Besides, washing machine effluent was also a significant source of microplastic here. Their study showed higher microplastic abundance in the area which displayed higher antibiotic resistance.

3.3 Lacustrine Environment

Water bodies, including lakes and estuaries, were found to be both source and sink of microplastics [120]. Other than urban lakes, microplastics are also reported from lakes in remote areas [34]. The small-sized plastics have also been reported from the organisms living in these water bodies. In India, the first study on a lake was conducted by Sruthy and Ramaswamy [118]. They found out that the Vembanad Lake, a Ramsar site in India, is heavily polluted with Mps. A higher abundance of microplastics was observed from marine-influenced sites, and the microplastics were found to be discharged from nearby cities like Kochi and Eloor. The study proved that lake sediments act as a sink for these micro pollutants. They also highlighted the significance of residence time of microplastics in water which will enhance the biofouling and adsorption of various chemicals to the microplastic surface. Later, [39], examined the presence of microplastics in a lake in Chennai, Tamil Nadu—Lake Red Hills. They have observed microplastic contamination in the sediments and water samples from the lake. Fishing activities and dry deposition can be sources of these plastic materials. Besides, the sewage waste flow from the residential area close to the lake is a major contributor to the pollution in the lake. The authors show doubt about the efficiency of the water facility in removing the microplastics. They also showed their concern about the harmful effect of drinking the boiled water containing Mps, as the toxicity may increase as boiling results in the release of multiple metals present on the surface of the microplastics into the water. SEM study revealed the presence of various metals on the surface of Mps. Veeranam Lake of Tamil Nadu is also polluted with microplastics with about more than 60% of the obtained microplastics were lesser than 1 mm in size [12]. The authors found out that the primary source of these materials in the lake is riverine input.

3.4 Groundwater

Microplastics contamination in groundwater is a severe global issue which needs to be adequately tackled. In India, a few studies have been reported from groundwater. Selvam et al. [110] investigated microplastic distribution in the surface and subsurface water from coastal southern India. They have linked the higher abundance of Mps to industrial and tourist activities. Besides, they have analysed the adsorption capacities of various polymers for different pollutants. They have also suggested the OSPRC framework (Origins, sources, pathways, receptors, and consequences) for a complete representation of microplastic research. Ganesan et al. [36] analysed the microplastic abundance in groundwater along with various water sources from Chennai, including surface water and commercial drinking water. They were the first in India to study the occurrence of drinking water. They attributed the reason for the microplastic presence in the water to be the over usage of plastics and the absence of a proper waste management system.

3.5 Biota

The Mps research in India is in its emerging stage, and there are immense gaps of knowledge regarding freshwater Mps. There is very little data on the environmental fate, and the biological effects of Mps in freshwater species are unavailable. Intake of fishes can be a possible source of human intake of microplastics [24], especially consumption of dried fish (the whole body of the fish is dried; in most cases without removing the digestive tract; [23, 61]. As mentioned earlier, only a few reports were available on the presence of Mps in the organisms. Microplastics were detected from fish [60, 68], anchovies Kripa et al. [66], benthic invertebrates [81], oysters [89], mussels [80], etc. Naidu and his team confirmed the presence of Mps in the benthic invertebrates off Kochi sediments [81]. They provided the first evidence for the existence of microplastics particles in benthic fauna from the coastal waters of India. Microplastics (PE, CL, RY, PES, and PP) were obtained from important fishes collected from different coasts of Kerala [98]. A study conducted by James et al. [51] found 4.6% of microplastics from the fishes (n = 653; 16 species) collected by them from off Kochi. A shrimp species from the coastal waters of Kochi, *Fenneropenaeus indicus*, has been investigated for the presence of microplastics and their seasonal variation in them (mean $\pm$ SD = 0.39 $\pm$ 0.6 microplastics/shrimp; [23]). The result showed a comparatively higher abundance of microplastic pollution during the monsoon season. The study conducted by Daniel et al. [24] observed a significantly higher microplastic concentration in inedible tissue in filter feeders compared to visual predators. Their study obtained microplastic abundance in both edible and inedible parts of some commercially important pelagic fishes from the Kochi coast (n = 270; 30 fish per species), but in lower quantity. Most of the fishes had microplastics in the inedible parts (0.53 $\pm$ 0.77 items/fish) than in the edible parts (0.07 $\pm$ 0.26 items/fish).

Kumar et al. [68] found out the presence of microplastics in twelve fishes out of forty fishes collected Tuticorin, in which microfibres were dominant. As the coast is nearer to the Gulf of Mannar, the authors emphasized the need for particular concern to the region. Microplastics were obtained from the Indian edible oyster (*Magallana bilineata*; n = 180) collected from three locations along this coast (mean abundance = 6.9 $\pm$ 3.84 items/individual; [89]. They observed a strong relation between microplastic distribution along the oysters and water samples collected from the study area. Their results showed that microplastic bioavailability increases with the increasing size of oysters. Karuppasamy et al. [61] found out the presence of plastic particles (micro-and meso-plastics) from the gastrointestinal tract of some important fishes they collected from Chennai and Nagapattinam, Tamil Nadu.

After a study conducted on two bivalve species collected from the three estuaries in Pondicherry, the authors calculated the amount of ingestion of microplastics per year through mussel consumption by an average person belonging to the local community to be 3917.79 $\pm$ 144.71 [30]. The microplastic ingestion in both mesopelagic and epipelagic fishes was reported from Tuticorin, with higher levels of microplastic pollution, observed in the former compared to the latter. The microplastic concentration in the seawater collected from the coast revealed the influence of microplastic abundance on the concentration of Mps in fishes [107]. Apart from sediments and seawater samples, as mentioned in the earlier section, [40] studied microplastic pollution in some important organisms from Port Blair Bay. They observed high bioaccumulation of microplastics in zooplankton and other significant carnivorous fishes.

3.6 Salt

Microplastics were obtained from various salt brands produced in India. Seth and Shriwastav [112] found out that samples from all the salt brands (n = 8) of investigated sea salts were found contaminated, of which about 80% of the obtained plastics were of the size range of 2 mm and 0.5 mm. Salt samples (n = 25) were collected from salts pans in Tuticorin coastal region and investigated for microplastic presence Selvam et al. [111]. About 60% of the total pollutants were of size less than 0.1 mm. A study conducted on food-grade salts observed a higher abundance of Mps in the salts produced from seawater compared to that from borewell water [106]. Further, they observed the association of different environmental pollutants with Mps. An average of about 700 particles/kg of Mps was obtained from selected Indian commercial salts (n = 10; Sivagami et al. [114]. A higher abundance of microplastics was of size less than 0.5 mm. they also observed the effect of Mps in the salt on humans and found out the evidence of cell detachment on the exposure of Mps to the organs (tested on human embryonic kidney). Table salt can be a source of microplastic ingestion by humans, and this has to be resolved. Sand filtration of the seawater can be a method to remove Mps and thereby to reduce the entry of Mps into the salt [112].

3.7 Dust/Atmospheric Deposition

Atmospheric deposition is a significant way of transfer of microplastics through different parts ecosystems. In India, some researchers have investigated the amount of microplastics in the dust samples. The road dust samples from Nagpur—both urban and rural areas—have been examined for microplastic presence by Narmadha et al. [83]. One of the significant sources of microplastic pollution that they have observed in the study is the fibres released during washing and drying of laundry; cellulose fibres were dominant in the residential area. A similar study has been reported from Chennai; the road dust (n = 16) of this metropolitan city has been collected for microplastic investigation [88]. In the study, they observed that the south Chennai, a highly industrial area (265.42 ± 76.76 particles/100 g) showed comparatively higher average microplastic abundance than the northern part, a residential location

(190.46 ± 83.38 particles/100 g). A study by Zhang et al. [141] obtained different microplastic particles (TPA, PET, BPA, PC) from various indoor dust samples (n = 33) collected from different houses of Patna during 2014.

4 Pakistan

According to the International Trade Administration (2019), plastics account for around 6% of the total solid waste produced in Pakistan. Improper waste management practices often lead them to ending up in water bodies primarily rivers, and lakes. Irfan et al. [48] investigated the microplastic distribution in the River Ravi, Lahore. A mean concentration of 2074 ± 3651 Mps/m^3 and 3726 ± 9030 Mps/m^2 was found in the water and sediment samples, respectively. They observed microplastic distribution along a gradient of human impact showing municipal sewage as the greatest contributor [5, 16]. They reported a high proportion of large-sized (300 μm–5 mm) microplastics in the surface water samples compared to sediment samples unlike many other studies. This possibly could be due to the collection of samples near to industrial sources or due to insufficient fragmentation time [87]. Another study on the freshwater system in Pakistan was conducted by Irfan et al. [48]. They collected surface water and sediment samples from the Lake Rawal. They reported mean concentration of microplastics in water and sediments as 0.142 items/0.1 L and 1.04 items/0.01 kg, respectively. Fibres and fragments were the dominant categories of Mps found in this study which was also reported by Irfan et al. [49]. They observed a comparatively significant amount of microplastics in the bottom sediments than in the surface water similar to other studies [120, 137], possibly due to biofouling or due to low bottom currents in the lake.

Rafique et al. [92] investigated the impact of land-use practices and population density on the microplastic distribution in soils from Lahore. They observed an average concentration of microplastics around 4483 ± 2315 Mps/kg. High concentration of microplastics was found in the parks and lawns indicating the human influence on the microplastic concentration. Zhang et al. [141] examined the microplastics present in indoor dust samples from a dozen or so countries in which Pakistan was one of them. They quantitatively measured the freely available TPA and BPA present in the dust samples. PET-based microplastics were the major contributors showing a positive correlation between TPA and PET concentration [70, 77]. A significant correlation was observed between the GDP of the country and its indoor microplastic pollution. Concentrations of PET- and PC-based microplastics in urban area were higher than the rural area. However, they restricted the study to PET- and PC-based microplastics and there are many other sources which need to be addressed.

5 Maldives

Saliu et al. [103] studied the distribution of microplastic particles on beach sediments and surface water samples along the coral reef environment at Faafu Atoll, Maldives. They conducted a study on two different environments: (i) inner reef which is exposed to land-based inputs, and (ii) low-hydrodynamic conditions compared to the outer reef part. The variations were positively correlated with the microplastic abundance observed in the inner reef (0.31 particles/m^3) and outer reef 0.02 particles/m^3. However, long-distance transport of debris from the southern Indian Ocean [32] to the Maldives sea water [109] is also another reason cited as majority of the plastic particles collected showed severe degradation pattern suggesting long-distance travel before reaching the atoll. They also observed distribution of charred microplastics mainly due to the burning of plastic waste along shore lines. In another study, Saliu et al. [104] investigated the presence of phthalic acid esters (PAE)—a potent priority pollutant, causing adverse effects on the marine organisms [85, 86]. They identified a significant presence of PAE in scleractinian corals. Higher concentration of Mps was observed in the inner reef as stated in the previous study by Saliu et al. [103]. Imhoff et al. [47] investigated the plastic waste contamination along the shoreline sediments of Vavvaru, a remote coral island in Maldives. They estimated the long-term and daily abundance of plastic wastes along the shoreline. Their study pointed out the importance of concentrating more on the drifting plastics on ocean bodies as they found a significant daily abundance (35.8 ± 42.5 plastic particles/m2) of microplastics in the remote island like Vavvaru. They also observed high long-term accumulation (1029 ± 1134 plastic particles/m2) of microplastics in the similar environment suggesting the global distribution of microplastics from ocean gyres or neighbouring islands with higher population density. The shape and pattern of the plastics and plastic pellets presence suggested the long-term drifting of plastics through water bodies. Another study done by Patti et al. [90] investigated the sediment samples along Naifaru island in Maldives and estimated one of the highest densities of microplastics compared with the other observations worldwide. They observed around 55–1127.5 microplastics/kg. They suggested that the large-scale land reclamation happening along the coastal lines of Maldives influences the deposition of microplastics in the sediment samples from the marine depositional zones. This study also shows the long range transport of microplastics through marine environment.

6 Bangladesh

Microplastic pollution study in Bangladesh is still in the nascent stage. A study report (ESDO, 2016) by Environment and Social Development organization in 2015 was first of a kind dealing with microplastic pollution. Their study revealed the release of around 8000 million microbeads every month from cosmetics and cleaning products among three urban communities to the nearby water bodies. Hossain et al. [46] studied the intake of microplastic by three commercial marine fishes from Bangladesh coast. They observed higher abundance of microplastics in gastrointestinal tract of fishes with higher body weight suggesting a positive relationship between the two parameters. The larger size microplastics were observed in Bombay-duck species. They feed on shrimps and smaller fishes [7]. Bombay-duck is normally consumed by people without processing and the removal of digestive tract does not take place. This results in the bioaccumulation of microplastics to higher trophic level. The black [10], white colour [14] of the microplastics observed was in agreement to the feeding habit of fishes mistaking it to their natural prey [21]. A similar study was done by Hossain et al. [46] on the tiger shrimp and brown shrimp species. Larger microplastics were found on the tiger shrimp species compared to brown shrimp suggesting the variation in feeding habits and habitat differences between near shore and offshore regions [19]. Rahman et al. [93] studied the occurrence of microplastics and their distribution along the beach sediments in Cox's bazar, Bangladesh. They observed the mean abundance of microplastics in the site to be 8.1 ± 2.9 particles kg $^{-1}$ which is much lesser than reported elsewhere in the world. Their usage of mesh size of 300 μm could have resulted in the lower recovery of smaller size plastic in the laboratory analysis.

7 Sri Lanka

Jang et al. [52] published first of its kind study on marine debris along the coastal beaches of Srilanka. This study showed the significant contribution of plastic waste (98%) in the debris collected. However, their study was limited to mesoplastic range. There has been no significant research done in the case of microplastic in Srilanka though Balasubramanim and Phillott [6] compared the microplastic generated on Srilankan beaches to that of nine other countries and showed the abundance of microplastics around 10 Mps per 25 g of dry sediment considered. However, no characterization study was done. Koongolla et al. [65] studied the spatial distribution of microplastics in the beach sediments and surface water along the southern coastline of Srilanka. They observed significant amount of microplastics in both water (0–29 items/m2) and sediment samples (0-157 items/m2) collected from multiple locations. They compared the GDP and microplastic pollution of Srilanka with various other countries as plastic waste increases with GDP of a country [50]. The study showed similar results with that of Romania [91] and Morocco [58] though GDP is comparable only in the latter. However, such comparison is not dependable due to the differences in sampling and techniques employed. They observed high amount of Polypropylene and Polystyrene suggesting the sources like fishing nets, bags, and bottles. Similarily Polystyrene foam was found abundant mainly emitted from fish storage boxes and buoys. This is indicative of poor waste management in ports. Fragments, where more predominant shape observed, were compared to the pellets suggesting fragmentation of bigger plastics to smaller sizes. Carbonyl absorption spectra through FTIR-ATR however showed intensity values out of the normal absorption band (1715–1775 cm^{-1}) [31] suggesting the micoplastics were relatively new. This possibly could be due to the presence of UV stabilizers present in the Mps or the better waste management practices.

8 Nepal

Yukioka et al. [138] studied the microplastic presence in road dust samples in Kathmandu, Nepal. They found 12.5 ± 10.1 pieces/m2 of microplastic distributed in the samples collected. They observed microplastics from 100 μm-5 mm mainly from container or packaging Mps however they suggested the importance of identifying smaller microplastics <100 μm mainly from clothing fibres and tires [45, 132] that was not included in the study.

9 Recommendations for Future Studies

In the recently concluded decade, there have been several studies on microplastics and its impacts on the biota. The information derived from them has helped the community to refine the knowledge on microplastics and its behaviour in different ecosystems. When compared with the other countries, Mps research in SAARC nations is pretty much in the nascent stage of development. This is evident from the microplastic studies reported by researchers from SAARC nations. Apart from India, other countries such as Bangladesh, Sri Lanka, Pakistan, Nepal, and Maldives have not carried out any large-scale investigation on these tiny plastic particles and its impact on the biota. Hence, there are still several fundamental issues and serious questions which need to be immediately addressed. The Indian studies highlighted above are greatly restricted to the coastal and lacustrine environments and very few studies have been made on the major and minor rivers debouching into the Arabian Sea and the Bay of Bengal. This is important as rivers are considered to be one of the dominant pathways for the transfer of microplastics from the hinterland into the oceans/seas. Besides, no significant studies on microplastics have been made on the different types of soil samples (agricultural, urban, rural, forest, coastal, etc.) and their effects on the soil organisms. The vertical distribution of microplastics both in soils and sediments needs to be well studied. Microplastic investigation in soil profiles and sediment cores provides knowledge about the infiltration capacity of microplastics. Besides, it will help in understanding the effect of microplastics on various properties of soil and sediments such as permeability, porosity, and other mechanical characteristics. Further, mangrove sediments which are considered to be one of the major sinks of microplastics has been least studied in the SAARC nations.

It is still not known if the wastewater treatment plants of the SAARC countries are efficient enough to remove the microplastics of different types and sizes from the sewage water. Some studies have related microplastic distribution with water quality parameters [62, 105] and grain size of sediments [26]. More studies should focus on the inflow and fate of microplastics and their relationship with the above-mentioned parameters. Most importantly, the impact of microplastics on human health is not known from the context of the SAARC nations.

Further, in most of the studies, after the chemical procedures, Mps are identified using a stereo-zoom microscope and validated using FTIR techniques. A few studies have used Raman Spectroscopy for validation [118, 29, 30, 51, 80, 81, 88]. For precise mass estimation, future studies should make use of the pyrolysis GC-MS technique. Most of the studies show differences in field methodology within the same environment and laboratory analysis of microplastic particles making it difficult to compare the data. To minimize the errors, the global microplastics research community must devise a solid sampling strategy and analytical methods that can be used by researchers from different parts of the world and generate useful data on microplastic pollution.

10 Conclusion

It is a well-established fact that microplastics are omnipresent and its abundance in a particular ecosystem can have disastrous consequences on the biota. Most of the developed nations have moved ahead by leaps and bounds to study the minute details of the impact of microplastics on the terrestrial and aquatic ecosystems. However, in the SAARC nations, which are one of the largest plastic-producing countries, very little work has been made on environmental microplastic pollution. Except for India, the other SAARC countries have done very little in this field. Even in India, not many studies have been carried out to understand the microplastics flux rate into the Arabian Sea and Bay of Bengal. There is no information on the vertical migration of microplastics in the soil, sediment, and water and how microplastics behave under changing salinity conditions in tropical estuaries. Besides, microplastic distribution in the atmosphere is a largely unexplored area of research which needs to be taken up immediately. This is because the SAARC nations experience severe air pollution in different cities and microplastics like fibres are known to remain largely in suspension in the air column for a significantly long time. The need of the hour is for researchers from the SAARC nations should come forward and join hands to plan scientific investigations on the different microplastic issues discussed above. Conduct proper surveys in the different environmental samples (marine, lake, beach, river, air, biota), so that microplastic pollution in each of them can be quantified. The high-resolution data can then be used by policy decision makers to come out with suitable legislation measures that can protect our environment. A citizen-centric science project should be taken up so that citizens can also be made to act responsibly when it comes to adopt proper solid waste management practices. It is therefore time to not only reduce, reuse, and recycle plastics but also to refuse them such that in the decades to come the scale of microplastic pollution can be lowered.

References

1. Amrutha K, Warrier AK (2020) The first report on the source-to-sink characterization of microplastic pollution from a riverine environment in tropical India. Sci Total Environ 739:140377. https://doi.org/10.1016/j.scitotenv.2020.140377
2. Andrady LA (2011) Microplastics in the marine environment. Mar Pollut Bull 62(8):1596–1605. https://doi.org/10.1016/j.marpolbul.2011.05.030
3. Ashwini SK, Varghese GK (2019) Environmental forensic analysis of the microplastic pollution at Nattika Beach, Kerala. Environ Forensics 5922. https://doi.org/10.1080/15275922.2019.1693442
4. Athawuda AMGAD, Jayasiri HB, Thushari GGN et al (2020) Quantification and morphological characterization of plastic litter (0.30–100 mm) in surface waters ofoff Colombo, west coast of Sri Lanka. Environ Monit Assess 192:509. https://doi.org/10.1007/s10661-020-08472-2
5. Auta HS, Emenike CU, Fauziah SH (2017) Distribution and importance of microplastics in the marine environment: a review of the sources, fate, effects, and potential solutions. Environ Int 102:165–176. https://doi.org/10.1016/j.envint.2017.02.013
6. Balasubramaniam M, Phillott AD (2016) Preliminary observations of microplastics from beaches in the Indian ocean. Indian Ocean Turtle Newslett 23:13–16
7. Balli JJ, Chakraborty SK, Jaiswar AK (2006) Biology of Bombay duck, Harpodon nehereus (Ham., 1822), from Mumbai waters. India J Indian Fish Assoc 33:1–10
8. Barnes DKA, Galgani F, Thompson RC, Barlaz M (2009) Accumulation and fragmentation of plastic debris in global environments. Philos T R Soc B Biol Sci 364:1985–1998. https://doi.org/10.1098/rstb.2008.0205
9. Begum AHN, Kumar SS, Krishnan KG, Srinivasa M (2020) Distribution of microplastics in estuarine sediments along tamilnadu coast, Bay of Bengal, India. Int J Sci Res Eng Tr 6(2):814–821
10. Bellas J, Martínez-Armental J, Martínez-Cámara A, Besada V, Martínez-Gómez C (2016) Ingestion of microplastics by demersal fish from the Spanish Atlantic and Mediterranean coasts. Mar Pollut Bull 109(1):55–60. https://doi.org/10.1016/j.marpolbul.2016.06.026
11. Besseling E, Wegner A, Foekema EM, Van Den Heuvel-Greve MJ, Koelmans AA (2013) Effects of microplastic on fitness and PCB bioaccumulation by the lugworm Arenicola marina (L.). Environ Sci Technol 47(1):593–600. https://doi.org/10.1021/es302763x
12. Bharath KM, Srinivasalu S, Natesan U, Ayyamperumal R, Kalam NS, Anbalagan S et al (2021) Chemosphere Microplastics as an emerging threat to the freshwater ecosystems of Veeranam lake in south India: a multidimensional approach. Chemosphere 264:128502. https://doi.org/10.1016/j.chemosphere.2020.128502
13. Bhattacharya P, Lin S, Turner JP, Ke PC (2010) Physical adsorption of charged plastic nanoparticles affects algal photosynthesis. J Phys Chem C 114(39):16556–16561. https://doi.org/10.1021/jp1054759
14. Boerger CM, Lattin GL, Moore SL, Moore CJ (2010) Plastic ingestion by planktivorous fishes in the North Pacific Central Gyre. Mar Pollut Bull 60(12):2275–2278. https://doi.org/10.1016/j.marpolbul.2010.08.007
15. Boucher J, Friot D (2017) Primary microplastics in the Oceans: a global evaluation of sources. IUCN, Gland, Switzerland. https://doi.org/10.2305/IUCN.CH.2017.01.en

16. Browne MA, Crump P, Niven SJ, Teuten E, Tonkin A, Galloway T, Thompson R (2011) Accumulation of microplastic on shorelines woldwide: sources and sinks. Environ Sci Technol 45(21):9175–9179. https://doi.org/10.1021/es201811s
17. Centre for Coastal Zone Management and Coastal Shelter Belt (2020) http://iomenvis.nic.in/. Accessed 21 Nov 2020
18. Codina-García M, Militão T, Moreno J, González-Solís J (2013) Plastic debris in Mediterranean seabirds. Mar Pollut Bull 77(1–2):220–226. https://doi.org/10.1016/j.marpolbul.2013.10.002
19. Cole M, Lindeque PK, Fileman ES, Halsband C, Goodhead R, Moger J, Galloway TS (2013) Microplastic ingestion by zooplankton. Environ Sci Technol 47:6646–6655. https://doi.org/10.1021/es400663f
20. Cole M, Lindeque P, Halsband C, Galloway TS (2011) Microplastics as contaminants in the marine environment: a review. Mar Pollut Bull 62(12):2588–2597. https://doi.org/10.1016/j.marpolbul.2011.09.025
21. Collard F, Gilbert B, Eppe G, Roos L, Compere P, Das K, Parmentier E (2017) Morphology of the filtration apparatus of three planktivorous fishes and relation with ingested anthropogenic particles. Mar Pollut Bull 116:182–191. https://doi.org/10.1016/j.marpolbul.2016.12.067
22. Crawford CB, Quinn B (2017) Microplastic Pollutants, p 336. Elsevier: Amsterdam. https://doi.org/10.1016/C2015-0-04315-5
23. Daniel DB, Ashraf PM, Thomas SN (2020a) Abundance, characteristics and seasonal variation of microplastics in Indian white shrimps (Fenneropenaeus indicus) from coastal waters off Cochin, Kerala. India. Sci Total Environ 737:139839. https://doi.org/10.1016/j.scitotenv.2020.139839
24. Daniel DB, Ashraf PM, Thomas SN (2020b) Microplastics in the edible and inedible tissues of pelagic fishes sold for human consumption in Kerala. India. Environ Pollut 266:115365. https://doi.org/10.1016/j.envpol.2020.115365
25. Das P, Mohanty P (2016) Marine debris in little and great Nicobar Islands. J Aquacult Mar Bio 2017:8–10
26. Deng H, He J, Feng D, Zhao Y, Sun W, Yu H, Ge C (2021) Microplastics pollution in mangrove ecosystems: a critical review of current knowledge and future directions. Sci Total Environ 142041. https://doi.org/10.1016/j.scitotenv.2020.142041
27. Desforges JPW, Galbraith M, Dangerfield N, Ross PS (2014) Widespread distribution of microplastics in subsurface seawater in the NE Pacific Ocean. Mar Pollut Bull 79(1–2):94–99. https://doi.org/10.1016/j.marpolbul.2013.12.035
28. Dharani G, Abdul Nazar AK, Venkatesan R, Ravindran M (2003) Marine debris in Great Nicobar. Curr Sci 85:574–575
29. Dowarah K, Devipriya SP (2019) Microplastic prevalence in the beaches of Puducherry, India and its correlation with fishing and tourism / recreational activities. Mar Pollut Bull 148(June):123–133. https://doi.org/10.1016/j.marpolbul.2019.07.066
30. Dowarah K, Patchaiyappan A, Thirunavukkarasu C (2020) Quantification of microplastics using Nile Red in two bivalve species Perna viridis and Meretrix meretrix from three estuaries in Pondicherry, India and microplastic uptake by local communities through bivalve diet. Mar Pollut Bull 153(February):110982. https://doi.org/10.1016/j.marpolbul.2020.110982
31. Endo S, Takizawa R, Okuda K, Takada H, Chiba K, Kanehiro H et al (2005) Concentration of polychlorinated biphenyls (PCBs) in beached resin pellets: variability among individual particles and regional differences. Mar Pollut Bull 50(10):1103–1114. https://doi.org/10.1016/j.marpolbul.2005.04.030
32. Eriksen M, Lebreton LC, Carson HS, Thiel M, Moore CJ, Borerro JC (2014) Plastic pollution in the world's oceans: more than 5 trillion plastic pieces weighing over 250,000 tons afloat at sea. PLoS ONE 9(12):e111913. https://doi.org/10.1371/journal.pone.0111913
33. Fendall LS, Sewell MA (2009) Contributing to marine pollution by washing your face: microplastics in facial cleansers. Mar Pollut Bull 58(8):1225–1228. https://doi.org/10.1016/j.marpolbul.2009.04.025

34. Free CM, Jensen OP, Mason SA, Eriksen M, Williamson NJ, Boldgiv B (2014) High-levels of microplastic pollution in a large, remote, mountain lake. Mar Pollut Bull 85(1):156–163. https://doi.org/10.1016/j.marpolbul.2014.06.001
35. GESAMP (2015) Sources, fate and effects of microplastics in the marine environment: a global assessment: In: Kershaw PJ (ed) (IMO / FAO / UNESCO-IOC / UNIDO /WMO /IAEA /UN /UNEP/UNDP joint group of experts on the scientific aspects of marine environmental protection. Rep Stud GESAMP 90:96
36. Ganesan M, Nallathambi G, Srinivasalu S (2019) Fate and transport of microplastics from water sources. Curr Sci 00113891(11):117. https://doi.org/10.18520/cs/v117/i11/1874-1879
37. Ganesapandian S, Manikandan S, Kumaraguru AK (2011) Marine litter in the northern part of Gulf of Mannar, southeast coast of India. Res J Environ Sci 5(5):471. https://doi.org/10.3923/rjes.2011.471.478
38. González-Pleiter M, Pedrouzo-Rodríguez A, Verdú I, Leganés F, Marco E, Rosal R, Fernández-Piñas F (2020) Microplastics as vectors of the antibiotics azithromycin and clarithromycin: effects towards freshwater microalgae. Chemosphere 128824. https://doi.org/10.1016/j.chemosphere.2020.128824
39. Gopinath K, Seshachalam S, Neelavannan K, Anburaj V, Rachel M, Ravi S et al (2020) Quantification of microplastic in red hills lake of Chennai city, Tamil Nadu, India. Environ Sci Pollut Res 1–10. https://doi.org/10.1007/s11356-020-09622-2
40. Goswami P, Valsalan N, Dharani G (2020) First evidence of microplastics bioaccumulation by marine organisms in the Port Blair Bay. Andaman Islands. Mar Pollut Bull 155:111163. https://doi.org/10.1016/j.marpolbul.2020.111163
41. Guilhermino L, Vieira LR, Ribeiro D, Tavares AS, Cardoso V, Alves A, Almeida JM (2018) Uptake and effects of the antimicrobial florfenicol, microplastics and their mixtures on freshwater exotic invasive bivalve Corbicula fluminea. Sci Total Environ 622:1131–1142. https://doi.org/10.1016/j.scitotenv.2017.12.020
42. Harris PT (2020) The fate of microplastic in marine sedimentary environments: A review and synthesis. Mar Pollut Bull 158:111398. https://doi.org/10.1016/j.marpolbul.2020.111398
43. Haynes HM, Taylor KG, Rothwell J, Byrne P (2020) Characterisation of road-dust sediment in urban systems: a review of a global challenge. J Soil Sediment 1–24. https://doi.org/10.1007/s11368-020-02804-y
44. Hepsø MO (2018) Experimental weathering of microplastic under simulated environmental conditions-Method development and characterization of pristine, photodegraded and mechanically weathered microplastic (Master's thesis, NTNU)
45. Hernandez E, Nowack B, Mitrano DM (2017) Polyester textiles as a source of microplastics from households: a mechanistic study to understand microfiber release during washing. Environ Sci Technol 51(12):7036–7046. https://doi.org/10.1021/acs.est.7b01750
46. Hossain MS, Sobhan F, Uddin MN, Sharifuzzaman SM, Chowdhury SR, Sarker S, Chowdhury MSN (2019) Microplastics in fishes from the Northern Bay of Bengal. Sci Total Environ 690:821–830. https://doi.org/10.1016/j.scitotenv.2019.07.065
47. Imhof HK, Sigl R, Brauer E, Feyl S, Giesemann P, Klink S et al (2017) Spatial and temporal variation of macro-, meso-and microplastic abundance on a remote coral island of the Maldives. Indian Ocean. Mar Pollut Bull 116(1–2):340–347. https://doi.org/10.1016/j.marpolbul.2017.01.010
48. Irfan M, Qadir A, Mumtaz M, Ahmad SR (2020a) An unintended challenge of microplastic pollution in the urban surface water system of Lahore, Pakistan. Environ Sci Pollut Res 1–13. https://doi.org/10.1007/s11356-020-08114-7
49. Irfan T, Khalid S, Taneez M, Hashmi MZ (2020b) Plastic driven pollution in Pakistan: the first evidence of environmental exposure to microplastic in sediments and water of Rawal Lake. Environ Sci Pollut Res 1–10. https://doi.org/10.1007/s11356-020-07833-1
50. Jambeck JR, Geyer R, Wilcox C, Siegler TR, Perryman M, Andrady A, Narayan R, Law KL (2015) Plastic waste inputs from land into the ocean. Science 347(6223):768–771. https://doi.org/10.1126/science.1260352

51. James K, Vasant K, Padua S, Gopinath V, Abilash KS, Jeyabaskaran R, Babu A, John S (2020) An assessment of microplastics in the ecosystem and selected commercially important fishes o ff Kochi, south eastern Arabian Sea. India. Mar Pollut Bull 154:111027. https://doi.org/10.1016/j.marpolbul.2020.111027
52. Jang YC, Ranatunga RRMKP, Mok JY, Kim KS, Hong SY, Choi YR, Gunasekara AJM (2018) Composition and abundance of marine debris stranded on the beaches of Sri Lanka: results from the first island-wide survey. Mar Pollut Bull 128:126–131. https://doi.org/10.1016/j.marpolbul.2018.01.018
53. Jayasiri HB, Purushothaman CS, Vennila A (2015) Bimonthly variability of persistent organochlorines in plastic pellets from four beaches in Mumbai coast. India Environ Monit Assess 187:469. https://doi.org/10.1007/s10661-015-4531-5
54. Jayasiri HB, Purushothaman CS, Vennila A (2013a) Plastic litter accumulation on high-water strandline of urban beaches in Mumbai, India. Environ Monit Assess 185:7709–7719. https://doi.org/10.1007/s10661-013-3129-z
55. Jayasiri HB, Purushothaman CS, Vennila A (2013b) Quantitative analysis of plastic debris on recreational beaches in Mumbai. India Mar Pollut Bull 77(1–2):107–112. https://doi.org/10.1016/j.marpolbul.2013.10.024
56. Jeyasanta KI, Sathish N, Patterson J, Edward JKP (2020) Macro-, meso- and microplastic debris in the beaches of Tuticorin district, Southeast coast of India. Mar Pollut Bull 154:111055. https://doi.org/10.1016/j.marpolbul.2020.111055
57. Kaladharan P, Vijayakumaran K, Singh VV, Prema D, Asha PS, Sulochanan B (2017) Prevalence of marine litter along the Indian beaches : a preliminary account on its status and composition. J Mar Biol Ass India 59(1). https://doi.org/10.6024/jmbai.2017.59.1.1953-03
58. Kanhai LDK, Officer R, Lyashevska O, Thompson RC, O'Connor I (2017) Microplastic abundance, distribution and composition along a latitudinal gradient in the Atlantic Ocean. Mar Pollut Bull 115(1–2):307–314. https://doi.org/10.1016/j.marpolbul.2016.12.025
59. Kapinga CP, Chung SH (2020) Marine plastic pollution in South Asia. Development Papers 20–02, United Nations ESCAP, p 51
60. Karthik R, Robin RS, Purvaja R, Ganguly D, Anandavelu I, Raghuraman R et al (2018) Sci Total Environ Microplastics along the beaches of southeast coast of India. Sci Total Environ 645:1388–1399. https://doi.org/10.1016/j.scitotenv.2018.07.242
61. Karuppasamy PK, Ravi A, Vasudevan L, Prabu M (2020) Baseline survey of micro and mesoplastics in the gastro-intestinal tract of commercial fish from Southeast coast of the Bay of Bengal. Mar Pollut Bull 153:110974. https://doi.org/10.1016/j.marpolbul.2020.110974
62. Kataoka T, Nihei Y, Kudou K, Hinata H (2019) Assessment of the sources and inflow processes of microplastics in the river environments of Japan. Environ Pollut 244:958–965. https://doi.org/10.1016/j.envpol.2018.10.111
63. Kershaw DPJ (2016) Marine plastic debris and microplastics: global lessons and research to inspire action and guide policy change. Online document available at. United Nations Environment Programme. https://ec.europa.eu/environment/marine/good-environmental-status/descriptor10/pdf/Marine_plastic_debris_and_microplastic_technical_report_advance_copy.pdf
64. Klein S, Dimzon IK, Eubeler J, Knepper, TP (2018) Analysis, occurrence, and degradation of microplastics in the aqueous environment. In: Wagner M, Lambert S (eds) Freshwater Microplastics. The handbook of environmental chemistry, pp 51–67. Springer, Cham. https://doi.org/10.1007/978-3-319-61615-5_3
65. Koongolla JB, Andrady AL, Kumara PTP, Gangabadage CS (2018) Evidence of microplastics pollution in coastal beaches and waters in southern Sri Lanka. Mar Pollut Bull 137:277–284. https://doi.org/10.1016/j.marpolbul.2018.10.031
66. Kripa V, Nair, PG, Dhanya, AM, Pravitha, VP, Abhilash KS, Mohammed AA et al (2014) Microplastics in the gut of anchovies caught from the mud bank area of Alappuzha, Kerala. Marine Fish Inf Ser Tech Exten Ser 219:27–28
67. Krishnakumar S, Srinivasalu S, Saravanan P, Vidyasakar A, Magesh NS (2018) A preliminary study on coastal debris in Nallathanni Island, Gulf of Mannar Biosphere Reserve, Southeast coast of India. Mar Pollut Bull 131:547–551. https://doi.org/10.1016/j.marpolbul.2018.04.026

68. Kumar VE, Ravikumar G, Jeyasanta KI (2018) Occurrence of microplastics in fi shes from two landing sites in Tuticorin, South east coast of India. Mar Pollut Bull 135:889–894. https://doi.org/10.1016/j.marpolbul.2018.08.023
69. Lambert S, Wagner M (2018) Microplastics are contaminants of emerging concern in freshwater environments: an overview. In: Wagner M, Lambert S (eds) Freshwater Microplastics. The handbook of environmental chemistry, pp 58:1–23 Springer, Cham. https://doi.org/10.1007/978-3-319-61615-5_1
70. Le HH, Carlson EM, Chua JP, Belcher SM (2008) Bisphenol A is released from polycarbonate drinking bottles and mimics the neurotoxic actions of estrogen in developing cerebellar neurons. Toxicol Lett 176:149–156. https://doi.org/10.1016/j.toxlet.2007.11.001
71. Lee KW, Shim WJ, Kwon OY, Kang JH (2013) Size-dependent effects of micro polystyrene particles in the marine copepod Tigriopus japonicus. Environ Sci Technol 47(19):11278–11283. https://doi.org/10.1021/es401932b
72. Li J, Liu H, Chen JP (2018) Microplastics in freshwater systems: a review on occurrence, environmental effects, and methods for microplastics detection. Water Res 137:362–374. https://doi.org/10.1016/j.watres.2017.12.056
73. Maharana D, Saha M, Yousuf J, Rathore C, Sreepada RA, Xu X et al (2020) Assessment of micro and macroplastics along the west coast of India: abundance, distribution, polymer type and toxicity. Chemosphere 246:125708. https://doi.org/10.1016/j.chemosphere.2019.125708
74. Malizia A, Monmany-Garzia AC (2019) Terrestrial ecologists should stop ignoring plastic pollution in the Anthropocene time. Sci Total Environ 668:1025–1029. https://doi.org/10.1016/j.scitotenv.2019.03.044
75. Manickavasagam S, Kumar S, Kumar K, Bhuvaneswari GR, Paul T (2020) Quantitative assessment of influx and efflux of marine debris in a water channel of South Juhu creek, Mumbai. India. Reg. Stud. Mar. Sci. 34:101095. https://doi.org/10.1016/j.rsma.2020.101095
76. Masura J, Baker JE, Foster GD, Arthur C, Herring C (2015) Laboratory methods for the analysis of microplastics in the marine environment: recommendations for quantifying synthetic particles in waters and sediments. NOAA Technical Memorandum NOS-OR&R-48. National Oceanic and Atmospheric Administration, Silver Springs, MD, p 2015
77. Montuori P, Jover E, Morgantini M, Bayona JM, Triassi M (2008) Assessing human exposure to phthalic acid and phthalate esters from mineral water stored in polyethylene terephthalate and glass bottles. Food Addit Contam Part A 25:511–518. https://doi.org/10.1080/02652030701551800
78. von Moos N, Burkhardt-Holm P, Kohler A (2012) Uptake and effects of microplastics on cells and tissue of the blue mussel Mytilus edulis L. after an experimental exposure. Environ Sci Technol 46:11327–11335. https://doi.org/10.1021/es302332w
79. Mugilarasan M, Venkatachalapathy R, Sharmila N, Gurumoorthi K (2017) Occurrence of microplastic resin pellets from Chennai and Tinnakkara Island : towards the establishment of background level for plastic pollution. Indian J Mar Sci 46(06):1210–1212. http://nopr.niscair.res.in/handle/123456789/41992
80. Naidu SA (2019) Preliminary study and first evidence of presence of microplastics and colorants in green mussel, Perna viridis (Linnaeus, 1758), from southeast coast of India. Mar Pollut Bull 140:416–422. https://doi.org/10.1016/j.marpolbul.2019.01.024
81. Naidu SA, Rao VR, Ramu K (2018) Microplastics in the benthic invertebrates from the coastal waters of Kochi. South-eastern Arabian Sea. Environ Geochem Health 40(4):1377–1383. https://doi.org/10.1007/s10653-017-0062-z
82. Naik RK, Naik MM, D'Costa PM, Shaikh F (2019) Microplastics in ballast water as an emerging source and vector for harmful chemicals, antibiotics, metals, bacterial pathogens and HAB species: a potential risk to the marine environment and human health. Mar Pollut Bull 149:110525. https://doi.org/10.1016/j.marpolbul.2019.110525
83. Narmadha VV, Jose J, Patil S, Farooqui MO, Srimuruganandam B, Saravanadevi S, Krishnamurthi K (2020a) Assessment of Microplastics in Roadside Suspended Dust from Urban and Rural Environment of Nagpur, India. Int J Environ 1–12. https://doi.org/10.1007/s41742-020-00283-0

84. Narmadha VV, Jose J, Patil S, Farooqui MO, Srimuruganandam B, Saravanadevi S, Krishnamurthi K (2020b) Assessment of Microplastics in Roadside Suspended Dust from Urban and Rural Environment of Nagpur, India. Int J Environ 1–12. https://doi.org/10.1007/s41742-020-00283-0
85. Net S, Delmont A, Sempéré R, Paluselli A, Ouddane B (2015) Reliable quantification of phthalates in environmental matrices (air, water, sludge, sediment and soil): a review. Sci Total Environ 515–516:162–180. https://doi.org/10.1016/j.scitotenv.2015.02.013
86. Oehlmann J, Schulte-Oehlmann U, Kloas W, Jagnytsch O, Lutz I, Kusk KO et al (2009) A critical analysis of the biological impacts of plasticizers on wildlife. Philos Trans R Soc B 364:2047–2062. https://doi.org/10.1098/rstb.2008.0242
87. Pan Z, Guo H, Chen H, Wang S, Sun X, Zou Q, Zhang Y, Lin H, Cai S, Huang J (2019) Microplastics in the northwestern Pacific: abundance, distribution, and characteristics. Sci Total Environ 650:1913–1922. https://doi.org/10.1016/j.scitotenv.2018.09.244
88. Patchaiyappan A, Zaki S, Dowarah K (2020) Occurrence, distribution and composition of microplastics in the sediments of South Andaman beaches. Mar Pollut Bull 156:111227. https://doi.org/10.1016/j.marpolbul.2020.111227
89. Patterson J, Jeyasanta KI, Sathish N, Booth AM, Edward JKP (2019) Sci Total Environ Pro fi ling microplastics in the Indian edible oyster, Magallana bilineata collected from the Tuticorin coast, Gulf of Mannar, Southeastern India. Sci Total Environ 691:727–735. https://doi.org/10.1016/j.scitotenv.2019.07.063
90. Patti TB, Fobert EK, Reeves SE, da Silva KB (2020) Spatial distribution of microplastics around an inhabited coral island in the Maldives. Indian Ocean. Sci Total Environ 748:141263. https://doi.org/10.1016/j.scitotenv.2020.141263
91. Popa M, Morar D, Adrian T, Teuşdea A, Popa D (2014) Study concerning the pollution of the marine habitats with the microplastic fibres. J Environ Prot Ecol 15:916–923
92. Rafique A, Irfan M, Mumtaz M, Qadir A (2020) Spatial distribution of microplastics in soil with context to human activities: a case study from the urban centre. Environ Monit Assess 192(11):1–13. https://doi.org/10.1007/s10661-020-08641-3
93. Rahman SMA, Robin GS, Momotaj M, Uddin J, Siddique MAM (2020) Occurrence and spatial distribution of microplastics in beach sediments of Cox's Bazar. Bangladesh. Mar Pollut Bull 160:111587. https://doi.org/10.1016/j.marpolbul.2020.111587
94. Ram B, Kumar M (2020) Correlation appraisal of antibiotic resistance with fecal, metal and microplastic contamination in a tropical Indian river, lakes and sewage. NPJ Clean Water 3(1):1–12. https://doi.org/10.1038/s41545-020-0050-1
95. Rebolledo BEL, van Franeker JA, Jansen OE, Brasseur SM (2013) Plastic ingestion by harbour seals (Phoca vitulina) in The Netherlands. Mar Pollut Bull 67(1):200–202. https://doi.org/10.1016/j.marpolbul.2012.11.035
96. Reddy MS, Basha S, Adimurthy S, Ramachandraiah G (2006) Description of the small plastics fragments in marine sediments along the Alang-Sosiya ship-breaking yard, India. Estuar Coast Shelf Sci 68:656–660. https://doi.org/10.1016/j.ecss.2006.03.018
97. Rezania S, Park J, Din MFM, Taib SM, Talaiekhozani A, Yadav KK, Kamyab H (2018) Microplastics pollution in different aquatic environments and biota: A review of recent studies. Mar Pollut Bull 133:191–208. https://doi.org/10.1016/j.marpolbul.2018.05.022
98. Robin RS, Karthik R, Purvaja R, Ganguly D, Anandavelu I, Mugilarasan M, Ramesh R (2020) Holistic assessment of microplastics in various coastal environmental matrices, southwest coast of India. Sci Total Environ 703:134947. https://doi.org/10.1016/j.scitotenv.2019.134947
99. Rodrigues MO, Abrantes N, Gonçalves FJM, Nogueira H, Marques JC, Gonçalves AMM (2018) Spatial and temporal distribution of microplastics in water and sediments of a freshwater system (Antuã River, Portugal). Sci Total Environ 633:1549–1559. https://doi.org/10.1016/j.scitotenv.2018.03.233
100. Ryan PG, Moore CJ, van Franeker JA, Moloney CL (2009) Monitoring the abundance of plastic debris in the marine environment. Philos T R Soc B Biological Sciences 364(1526):1999–2012. https://doi.org/10.1098/rstb.2008.0207

101. Ryan PG (2015) A brief history of marine litter research. In: Bergmann M, Gutow L, Klages M (eds) Marine anthropogenic litter, pp 1–25. Springer: Cham. https://doi.org/10.1007/978-3-319-16510-3_1
102. Saha M, Veerasingam S, Suneel V, Naik BG, Vethamony P, Bhattacharya B (2018) Microplastic pollution on the coast of India. In: Bhattacharya B, de Boer J, Avino P (eds) Impact of pollutants of ecosystems and human health. Institute of Ecotoxicology and Environmental Sciences, Mudrakar, West Bengal, pp 155–166
103. Saliu F, Montano S, Garavaglia MG, Lasagni M, Seveso D, Galli P (2018) Microplastic and charred microplastic in the Faafu Atoll, Maldives. Mar Pollut Bull 136:464–471. https://doi.org/10.1016/j.marpolbul.2018.09.023
104. Saliu F, Montano S, Leoni B, Lasagni M, Galli P (2019) Microplastics as a threat to coral reef environments: detection of phthalate esters in neuston and scleractinian corals from the Faafu Atoll, Maldives. Mar Pollut Bull 142:234–241. https://doi.org/10.1016/j.marpolbul.2019.03.043
105. Sarkar DJ, Sarkar SD, Das BK, Manna RK, Behera BK, Samanta S (2019) Sci Total Environ Spatial distribution of meso and microplastics in the sediments of river Ganga at eastern India. Sci Total Environ 694:1–7. https://doi.org/10.1016/j.scitotenv.2019.133712
106. Sathish N, Jeyasanta KI, Patterson J (2019) Abundance, characteristics and surface degradation features of microplastics in beach sediments of five coastal areas in Tamil Nadu, India. Mar Pollut Bull 142:112–118. https://doi.org/10.1016/j.marpolbul.2019.03.037
107. Sathish MN, Jeyasanta I, Patterson J (2020) Sci Total environ occurrence of microplastics in epipelagic and mesopelagic fishes from Tuticorin, Southeast coast of India. Sci Total Environ 720:137614. https://doi.org/10.1016/j.scitotenv.2020.137614
108. Sathish MN, Jeyasanta I, Patterson J (2020a) Microplastics in sàlt of tuticorin, Southeast Coast of India. Arch Environ Contam Toxicol 1–11. https://doi.org/10.1007/s00244-020-00731-0
109. Van Sebille EV, Wilcox C, Lebreton L, Maximenko N, Hardesty BD, van Franeker JA, Eriksen M, Siegel D, Galgani F, Law KL (2015) A global inventory of small floating plastic debris. Environ Res Lett 10:124006. https://doi.org/10.1088/1748-9326/10/12/124006
110. Selvam S, Jesuraja K, Venkatramanan S, Roy PD, Kumari VJ (2020a) Hazardous microplastic characteristics and its role as a vector of heavy metal in groundwater and surface water of coastal south India. J Hazard Mater 402:123786. https://doi.org/10.1016/j.jhazmat.2020.123786
111. Selvam S, Manisha A, Venkatramanan S, Chung SY, Paramasivam CR, Singaraja C (2020b) Microplastic presence in commercial marine sea salts: a baseline study along Tuticorin Coastal salt pan stations, Gulf of Mannar. South India. Mar Pollut Bull 150:110675. https://doi.org/10.1016/j.marpolbul.2019.110675
112. Seth CK, Shriwastav A (2018) Contamination of Indian sea salts with microplastics and a potential prevention strategy. Environ Sci Pollut Res 25:30122–30131. https://doi.org/10.1007/s11356-018-3028-5
113. Sharma S, Chatterjee S (2017) Microplastic pollution, a threat to marine ecosystem and human health: a short review. Environ Sci Pollut Res 24:21530–21547
114. Sivagami M, Selvambigai M, Devan U, Velangani AAJ, Karmegam N, Biruntha M et al (2020a) Extraction of microplastics from commonly used sea salts in India and their toxicological evaluation. Chemosphere 263:128181. https://doi.org/10.1016/j.chemosphere.2020.128181
115. Sivagami M, Selvambigai M, Devan U, Velangani AAJ, Karmegam N, Biruntha M, Kim W, Govarthanan M, Kumar P (2020b) Jo ur na l P re. ECSN, 128181. https://doi.org/10.1016/j.chemosphere.2020.128181
116. Skalska K, Ockelford A, Ebdon JE, Cundy AB (2020) Riverine microplastics: Behaviour, spatio-temporal variability, and recommendations for standardised sampling and monitoring. J Water Process Eng 38:101600. https://doi.org/10.1016/j.jwpe.2020.101600
117. Sridhar KR, Deviprasad B, Karamchand KS, Bhat R (2009) Plastic debris along the beaches of karnataka, southwest coast of India. Asian J Water Environ Pollut 6(2):87–93

118. Sruthy S, Ramasamy EV (2017) Microplastic pollution in Vembanad Lake, Kerala, India: the first report of microplastics in lake and estuarine sediments in India. Environ Pollut 222:315–322. https://doi.org/10.1016/j.envpol.2016.12.038
119. Steer M, Cole M, Thompson RC, Lindeque PK (2017) Microplastic ingestion in fish larvae in the western english channel. Environ Pollut 226:250–259. https://doi.org/10.1016/j.envpol.2017.03.062
120. Su L, Xue Y, Li L, Yang D, Kolandhasamy P, Li D, Shi H (2016) Microplastics in Taihu lake, China. Environ Pollut 216:711–719. https://doi.org/10.1016/j.envpol.2016.06.036
121. Sulochanan B, Lavanya S, Kemparaju S (2013) Influence of river discharge on deposition of marine litter. Marine Fish Inf Ser T&E Ser 216:27–29
122. Sulochanan B, Bhat GS, Lavanya S, Dineshbabu AP, Kaladharan P (2014) A preliminary assessment of ecosystem process and marine litter in the beaches of Mangalore. Ind J Geo Mar Sci 43 (9):1764–1769. http://nopr.niscair.res.in/handle/123456789/34503
123. Sundar S, Chokkalingam L, Roy PD, Usha T (2020) Estimation of microplastics in sediments at the southernmost coast of India (Kanyakumari). Environ Sci Pollut Res 1–6. https://doi.org/10.1007/s11356-020-10333-x
124. Thompson RC, Moore CJ, VomSaal FS, Swan, SH (2009) Plastics, the environment and human health: current consensus and future trends. Phil Trans Roy Soc B Biol Sci 364(1526):2153–2166. https://doi.org/10.1098/rstb.2009.0053
125. Tirelli V, Suaria G, Lusher AL (2020) Microplastics in Polar Samples. In: Rocha-Santos T, Costa M, Mouneyrac C (eds) Handbook of Microplastics in the Environment. Springer, Cham. https://doi.org/10.1007/978-3-030-10618-8_4-1
126. Tiwari M, Rathod TD, Ajmal PY, Bhangare RC, Sahu SK (2019) Distribution and characterization of microplastics in beach sand from three different Indian coastal environments. Mar Pollut Bull 140:262–273. https://doi.org/10.1016/j.marpolbul.2019.01.055
127. Veerasingam S, Mugilarasan M, Venkatachalapathy R, Vethamony P (2016) Influence of 2015 flood on the distribution and occurrence of microplastic pellets along the Chennai coast, India. Mar Pollut Bull 109:196–204. https://doi.org/10.1016/j.marpolbul.2016.05.082
128. Veerasingam S, Saha M, Suneel V, Vethamony P, Rodrigues AC, Bhattacharyya S, Naik BG (2016) Characteristics, seasonal distribution and surface degradation features of microplastic pellets along the Goa coast, India. Chemosphere 159:496–505. https://doi.org/10.1016/j.chemosphere.2016.06.056
129. Vidyasakar A, Krishnakumar S, Kasilingam K, Neelavannan K, Bharathi VA, Godson PS et al (2020) Characterization and distribution of microplastics and plastic debris along Silver Beach. Southern India. Mar Pollut Bull 158:111421. https://doi.org/10.1016/j.marpolbul.2020.111421
130. Vidyasakar A, Neelavannan K, Krishnakumar S, Prabaharan G, Priyanka TSA, Magesh NS et al (2018) Macrodebris and microplastic distribution in the beaches of Rameswaram Coral Island, Gulf of Mannar, Southeast coast of India: a first report. Mar Pollut Bull 137:610–616. https://doi.org/10.1016/j.marpolbul.2018.11.007
131. Wagner M, Scherer C, Alvarez-Muñoz D, Brennholt N, Bourrain X, Buchinger S, Rodriguez-Mozaz S (2014) Microplastics in freshwater ecosystems: what we know and what we need to know. Environ Sci Eur 26(1):1–9. https://doi.org/10.1186/s12302-014-0012-7
132. Wang T, Zou X, Li B, Yao Y, Zang Z, Li Y, Yu W, Wang W (2019) Preliminary study of the source apportionment and diversity of microplastics: taking floating microplastics in the South China Sea as an example. Environ Pollut 245:965–974. https://doi.org/10.1016/j.envpol.2018.10.110
133. Wegner A, Besseling E, Foekema EM, Kamermans P, Koelmans AA (2012) Effects of nanopolystyrene on the feeding behavior of the blue mussel (Mytilus edulis L.). Environ Toxicol Chem 31(11):2490–2497. https://doi.org/10.1002/etc.1984
134. Whitmire SL, Van Bloem SJ (2017) Quantification of microplastics on national park beaches. NOAA Marine Debris Program. National Oceanic and Atmospheric Administration. https://marinedebris.noaa.gov/sites/default/files/publications-files/Quantification_of_Microplastics_on_National_Park_Beaches.pdf

135. Wright SL, Thompson RC Galloway TS (2013) The physical impacts of microplastics on marine organisms: a review. Environ Pollut 178:483492. https://doi.org/10.1016/j.envpol.2013.02.031
136. Xu JL, Thomas KV, Luo Z, Gowen AA (2019) FTIR and Raman imaging for microplastics analysis: state of the art, challenges and prospects. TRAC 119:115629. https://doi.org/10.1016/j.trac.2019.115629
137. Yaun W, Liu X, Wang W, Di M, Wang J (2019) Microplastic abundance, distribution and composition in water, sediments and wild fish from Poyang Lake, China. Ecotoxicol Environ Saf 170:180–187. https:// doi.org/https://doi.org/10.1016/j.ecoenv.2018.11.126
138. Yukioka S, Tanaka S, Nabetani Y, Suzuki Y, Ushijima T, Fujii S et al (2020) Occurrence and characteristics of microplastics in surface road dust in Kusatsu (Japan), Da Nang (Vietnam), and Kathmandu (Nepal). Environ Pollut 256:113447. https://doi.org/10.1016/j.envpol.2019.113447
139. Zettler ER, Mincer TJ, Amaral-Zettler LA (2013) Life in the "Plastisphere": microbial communities on plastic marine debris. Environ Sci Technol 47(13):7137–7146. https://doi.org/10.1021/es401288x
140. Zhang H (2017) Transport of microplastics in coastal seas. Estuar Coast Shelf Sci 199:74–86. https://doi.org/10.1016/j.ecss.2017.09.032
141. Zhang J, Wang L, Kannan K (2020) Microplastics in house dust from 12 countries and associated human exposure. Environ Int 105314. https://doi.org/10.1016/j.envint.2019.105314
142. Zitko V, Hanlon M (1991) Another source of pollution by plastics: skin cleansers with plastic scrubbers. Mar Pollut Bull 22:41–42

Distribution and Impact of Microplastics in the Aquatic Systems: A Review of Ecotoxicological Effects on Biota

Tadele Assefa Aragaw and Bassazin Ayalew Mekonnen

Abstract Microplastics (MPs) are emerging pollutants attracting attention due to there have been widely distributed in the aquatic and terrestrial environment. MPs have been quantified and identified in marine, freshwater, and terrestrial environments from biota samples. May originate from macroplastic waste due to improper management of them and also from the primary sources, so that emerges as a problem for the aquatic biota. This review aimed to discuss (a) the occurrence of MP, (b) the interaction and uptake mechanism, (c) effects on the biota, (d) microplastics and associated pollutant effect on the aquatic biota. The retrieved literature confirmed us MP and its associated contaminants have adverse effects on aquatic biota, thereafter affect human health through the food chain. Finally, future research directives and recommendations on the MP pollution research, emphasized on the experimental ecotoxicological studies and risk assessments of MP to aquatic organisms were highlighted.

Keywords Microplastics · Aquatic pollution · Aquatic biota · Microplastics ingestion · Interaction · Mechanism · Ecotoxicity · Associated contaminant

Abbreviations

AChE	Anti-cholinesterase
Ag	Silver
BAF	Bioaccumulation Factor

T. A. Aragaw (✉) · B. A. Mekonnen
Faculty of Chemical and Food Engineering, Bahir Dar Institute of Technology-Bahir Dar University, Bahir Dar, Ethiopia
e-mail: taaaad82@gmail.com

B. A. Mekonnen
e-mail: bassa.ched@gmail.com

B. A. Mekonnen
Bahir Dar Energy Center, Bahir Dar Institute of Technology-Bahir Dar University, Bahir Dar, Ethiopia

S. S. Muthu (ed.), *Microplastic Pollution*, Sustainable Textiles: Production, Processing, Manufacturing & Chemistry, https://doi.org/10.1007/978-981-16-0297-9_3

BPA	Bisphenol A
CAT	Catalase
Cu	Copper
DDT	Dichlorodiphenyltrichloroethane
EDC	Endocrine-Disrupting Chemicals
EROD	Ethoxyresorufin-O-Deethylase
FPs	Fine Plastic Particles
FTIR	Fourier Transform Infrared
FTIR-ATR	Fourier Transform Infrared- Attenuated Total Reflectance
HDL	High-density lipoprotein
HDPE	High-Density Polyethylene
Hg	Mercury
HOC	Hydrophobic organic contaminants
IDH	Isocitrate Dehydrogenase
LDH	Enzymes Lactate Dehydrogenase
LDPE	Low-Density Polyethylene,
LPO	Lipid Oxidation
MDA	Malondialdehyde
MPs	Microplastics
NOM	Natural organic matter
Ni	Nickel
NPs	Nanoparticles
PA	Polyamide
PAHs	Polycyclic Aromatic Hydrocarbons
PAN	Polyacrylonitrile
PAN	Polyacrylonitrile
PBDEs	Polybrominated Diphenyl ethers
PCBs	Polychlorinated Biphenyls

1 Introduction

Plastic production began in the early 1940s and was realized in 1950. The global plastic production surges from 2 million tons by 1950–381 million tons by 2015 [1] and is estimated to reach 33 billion tons by 2050 [2]. In the early 1940s, plastics production steered in a rapid increase in the amount being manufactured, leading to widespread utilization of plastics in near-unlimited applications [3]. As a result, plastics are vigorously used globally due to the low cost, ease of production, waterproofness, and longevity [4]. However, the major problem of plastics is their disposal and persistent accumulation in marine and freshwater environments [5, 6]. While most plastics are single-use, they are disposed of within one year of production, and the majority ends up in landfills [7]. For instance, the majority of overall worldwide plastic waste (55%) was discarded as either littered or defectively predisposed (mismanaged) in the year 1950 through 2015 [1]. In addition to the conventional plastic products (bags, bottles, and others), plastic nanofibers from medical areas have been huge MPs sources to the marine environment. For example surgical face masks, hand-on gloves, and other personal protective equipment [8]. The inappropriate dumping of plastic causes 80% of plastic contamination drives from land-based origins with the remainder originates from ocean-based sources of plastic on the aquatic system [9–11]. For example, an anticipated quantity of 4.8–12.7 Mt of mishandled plastic waste is supposed to have been left in the ocean by 2010 [11, 12]. From this plastic waste, an estimated quantity of over 80% of plastic fragments has been disposed into oceans mainly from 20 coastlines [11]. In the forthcoming, mismanaged plastic waste continued and projected in East Asia and Pacific (60%), South Asia (12%), Sub-Saharan Africa (11%), Middle East and North Africa (7.7%), Latin America and the Caribbean (6.3%), Europe and Central Asia (2.6%), and North America (0.5%) by 2025 [11]. The huge plastic production volume together with their high durability led to an extensive collection of discarded plastic wastes in landfills and plastic litter in terrestrial and water body surroundings globally [13]. This plastics debris progressively fragments into smaller pieces and becomes a source of microplastics (MP) contamination [7]. As a result, macroplastics and MPs are persistent contaminants that are being perceived as littering ecological niche of the globe [11, 14]. Particularly, MPs, which have been reported sporadically early the 1970s and estimated quantities over 5 trillion MPs are suspended in the ocean water bodies from 2007 to 2013 [15]. Currently, MPs are believed to be the predominant form of discharged plastic waste [16]. This is due to the marginal emphasis given to the small-sized plastic product management at the early stages. In addition, plastic waste disposal departing aside the effect of degradation under the ecosystem and ends up with the formation of smaller plastic particles. Based on the size, MPs are categories into four as MPs (<5 mm), mesoplastics (5–50 mm), macroplastics (50–500 mm), and macroplastics (>500 mm) [17]. Basically, <5 mm size of plastic fragments are eventually termed as MPs [6, 18].

Anthropogenic activity is the main cause of MPs pollution evidenced in several ecologies from terrestrial to aquatic habitats. The need for accurate assessment of the level of MPs in aquatic biota is crucial for determining baseline levels of contamination and assessing the risk of MP to species assemblage and ecosystem.

Given the important ecological role of freshwater and marine biota and the implications of microplastics in the ecosystem, this study aimed to assess and summarizes the occurrence, interaction, and ecotoxicity of MPs and associated heavy metals and organic contaminants to the range of freshwater and marine organisms. In addition, future research directives and recommendations in this review are provided.

2 Review Methods and Data Treatment

A comprehensive literature review of published articles was retrieved using the ISI Web of Science, Scopus, PubMed database, and other research-researcher networks (Google Scholar, Mendeley, and ResearchGate) for plastics and MPs study published up to September 2020. The keyword queries included 'Microplastics, 'microplastics occurrence' in a combination of 'biota', 'ingestion/uptake/transfer', and 'effects/impacts/toxicity'. The retrieved publications were then previewed separately to eliminate replicates, irrelevant, and unrelated literature. Furthermore, the retrieved published materials, such as research articles, review articles, case studies, technical reports, short communications, and books were downloaded. Further, the downloaded documents were screened by type of the studied water body (freshwater and marine water), biological groups of studied biota, the observed occurrence of MPs in biota, the observed ecotoxicological effects of MPs, the detected ecotoxicological effects of combined MPs, and associated pollutants with respective of countries. Then, published materials are categorized based on research scopes such as the occurrence of MPs in biota studies, ecotoxicity observed in the global aquatic system including across countries such as rivers, lakes, oceans, and gulfs and bay were selected. The impact of MPs associated with other contaminants for the range of fauna and flora from global water systems was also included.

3 The Occurrence of Microplastics in Aquatic Biota

MPs (<5 mm) have been discovered all over the world in a diversity of aquatic biota, comprising bivalves, birds, and fish. Several laboratory investigations have also revealed the existence of MPs in biota through direct ingestion or trophic transfer from contaminated prey to a range of trophic levels. Hence, MPs have been increasingly perceived and quantified in marine and freshwater environments and become highly bioavailable to marine and freshwater organisms [16, 19, 20].

For instance, a recent study by Sathish et al. (2020) assessed the incidence of MPs ingestion in seawater fish species samples from different habitats of Thirespuram coastal water in Tuticorin, southeast coast of India (Table 1). According to the authors [21], MPs <500 μm were identified in the fish gastrointestinal tracts with a concentration range of 0.11 ± 0.06–3.64 ± 1.7 items/individual fish which is equivalent to 0.0002 ± 0.0001–0.2 ± 0.03 items/g gut weight in the samples. In addition, FTIR-ATR analysis shown the different shapes of MPs including fiber, fragment, film, and foam in the samples of water and fish species. As confirmed by FTIR-ATR, small-sized (<500 μm), fibers, and blue colored PE were the dominantly found plastics in water and fish. Regardless of the plastic polymer identity found in the fishes, PE was the most frequently recorded MP, followed by PS and polyamide. Furthermore, the abundance of MPs in fishes is a function of the concentration levels of MPs in the Thirespuram coastal water environment. For example, the

Table 1 Abundance of MPs in biota species from marine and freshwater system

Water body	Biota species	Location	MPs concentration or occurrence/Abundance	MPs size	Polymer type; shape; color	Country	References
Marine	*Harpodon nehereus, Chirocentrus dorab, Sardinella albella, Rastrelliger kanagurta, Katsuwonus pelamis. and Istiophorus platypterus*	Tuticorin, southeast coast of India	0.11 ± 0.06 to 3.64 ± 1.7 items/individual 0.0002 ± 0.0001 to 0.2 ± 0.03 items/g gut weight	<500-µm	PE, PA; fiber type; blue color	India	[21]
Marine	*Rastrilleger kanagurta and Epinephalus merra*	Thirespuram and Punnakayal fish landing sites at Tuticorin, Southeast coast of India	30% of fish sample contaminated by MPs	0.1–0.5 mm	PE, PP; 80%microfibers, 20% fragment; red, black, and translucent colors	India	[22]
Marine	*Lateolabrax maculate (Fish)*	Hangzhou Bay and Yangtze Estuary	0.3 to 5.3 items/individual (i.e., 0.1 to 8.8 items/g) in gut, 0.3 to 2.6 items/individual. (i.e., 0.1 to 5.2 items/g) in gill	> 20-µm	PE.PP,PES, non-plastic (Cotton); Fibrous characteristics	China	[23]
Marine	*Scrobicularia plana*	Sigma-Aldrich	1 mg/L	20-µm	PS	Germany	[24]
Marine	Fish larvae	Western English Channel, offshore Plymouth	2.9% of fish larvae had ingested MPs	100–250-µm	Nylon, Polyamide-polypropylene, Rayon, Unknown polymer with 66% fiber, and 34% fragment; 84% blue, 16% red in color	UK	[25]

(continued)

Table 1 (continued)

Water body	Biota species	Location	MPs concentration or occurrence/Abundance	MPs size	Polymer type; shape; color	Country	References
Marine	*Neocalanus cristatus* *Euphausia pacifia*	Northeast Pacific Ocean	1 particle per 34 copepods 1 particle per 17 euphausiids	555.5 ± 148.7-µm 816.1 ± 107.7-µm	Polystyrene spheres, 44% fibers, 56% fragments in Cristatus Polystyrene spheres, 68% fibers, 32% fragments Pacifia	British Columbia, Canada	[26]
Marine	*bmussel Mytilus edulis and lugworm Arenicola marina*	French-Belgian-Dutch North Sea coast	On average: 0.2 ± 0.3 particles/g tissue On average: 1.2 ± 2.8 particles/g (0.2 ± 0.3	<1 mm	LDPE, HDPE and PS; blue	Frenche-Belgiane-Dutch coastline	[27]
Marine	*Mytilus edulis* *Crassostrea gigas*	The North Sea and the Atlantic Ocean	0.36 ± 0.07 particles/g (wet weight) 0.47 ± 0.16 particles/g (wet weight)	<1 mm	Red particle, Green sphere	Germany	[28]
Marine	*Galeus melastomus*	Western Mediterranean Sea	0.34 ± 0.07 particles per individual	<5 mm	Cellophane (33.33%), Polyacrylonitrile (4.55%), PE (4.55%), PET (4.55%), Poly Ethyl Acrylate (27.27%), Polyacrylate (1.52%), Polyamide (PA) (12.12%), Polypropylene (PP) (3.03%), Aklyd (1.52%). Containing filament (86.36%), fragment (12.12%), and film (1.51%) polymer; Black, Blue, Red, Transparent, White color	Balearic Islands	[29]
Marine	*Loggerhead sea turtles*	South-West Indian Ocean	96.2% of the turtle-ingested debris 79.88% ± 3.69 per turtle Hard plastic 11.29% ± 2.65 per turtle soft plastic 2.94 ± 0.79 per turtle plastics cap	> 0.5 cm debris 0.1 cm fragments	Hard plastic, Soft plastic, Plastic caps; hard white and blue	Between Madagascar and Reunion Island	[30]
Marine	*Megaptera novaeanglia*	Johanna	A total of 16 plastic particles	1 mm–17 cm	PE, PP, PVC, PET, nylon; sheets, fragments, and threads	Netherlands	[31]

(continued)

Table 1 (continued)

Water body	Biota species	Location	MPs concentration or occurrence/Abundance	MPs size	Polymer type; shape; color	Country	References
Marine	*Phoca vitulina*	Island Texel	0.26 particles per individual	0.12–0.3 mm	–	Netherlands	[32]
Freshwater	macroinvertebrate	Upstream and downstream Wastewater Treatment Works (WwTWs)	Approximately 50% of macroinvertebrate contains MPs 0.14 MP mg/tissue	100 μm	Fibers polymer	UK	[33]
Freshwater	*Gambusia holbrooki*	Greater metropolitan Melbourne wetlands	0.60 item/individual 19.4% of the fish collected contaminated	0.09–4.86	PS (25.7%), rayon (10.1%), PA (7.3%), and PP (5.5%) In particular, PS and rayon occurred in more than 70% of the sampling site.	Australia	[34]
Freshwater	*Carassius auratus*	Poyang Lake	91% of contaminated individuals 0–18 items per individual	< 0.5 mm	PP and PE; fibrous MPs;	China	[35]
Freshwater	*Gymnocypris przewalskii*	Qinghai Lake	5.4 item/individual	0.1–0.5 mm	PE and PP; sheet and fiber shapes	China	[36]
Freshwater	*Dicentrarchus labrax, Diplodus Vulgaris (Platichthys flesus)*	Mondego River Estuaries	1.67 ± 0.27 (SD) per fish	1–5 mm	PS, PP. and rayon; (96%) fibers, fragments (4%); blue color (47%), transparent (30%), and black (11%)	Portugal	[37]
Freshwater	Riverine fish	Lake Michigan tributaries	10 (± 2.3) to 13 (± 1.6) MPs particles/fish	<1.5 mm	97–100% fibers MPs in the water and fish, 1.5–3% fragment; blue fibers	USA	[17]
Freshwater	*Squalius cephalus*	arne & Seine Rivers	Mean 0.36 Anthropogenic particles/g SC (wet weight	Mean 2.00 mm–2.93 mm	PP, PET fiber, fiber, fragment,	France	[38]
Freshwater	Hoplosternum littorale	Pajeú River	83% of the fish ingested plastic debris with 3.6 particles/fish	<5 mm	Plastic debris; Fibers, soft, hard	Brazil	[39]
Freshwater	Coastal freshwater fish	The Río de la Plata (RLP) estuary (Southern Coastal Fringe)	18.5 ± 18.9 fibers/fish 0.7 ± 1.7 'others'/fish	0.06–4.7 mm	Plastic debris; 96% fibers, and 4% to 'others'	Argentina	[40]

(continued)

Table 1 (continued)

Water body	Biota species	Location	MPs concentration or occurrence/Abundance	MPs size	Polymer type; shape; color	Country	References
Freshwater	*Tubifex tubifex worms*	River Irwell, England	129 ± 65.4 particles/g tissue (wet weight)	50–4500 μm in length	PET, PS and acrylic fibers; fibers (87%), 13% Fragments; typically blue (50%), black (22%), or red (9%) color	UK	[41]
Freshwater	*Corbicula flumine*	Yangtze River	0.3–4.9 items/g (or 0.4–5.0 items/individual)	0.021–4.83 mm 0.25–1 mm dominant	PS(33%), followed by PP (19%) and PE (9%); 60–40% fiber, 30% Blue and transparent color, 70% other colors (blue, red, green, yellow, white)	China	[42]
Freshwater	*Gymnocypris przewalskii*	Xiangxi River,	25.7% of the collected fish contained microplastics	0.3−1.8	PE in single fragment recognized as nylon. Other particles are recognized as carbonate, wood, pigment, and quartz.	China	[43]
Freshwater	*Ardea cinerea, Cygnus olor Anas platyrhynchos*	Lake Geneva,	4.3 ± 2.6 particles per bird	<5 mm	62% PE, 15% PP, 12% PS, 4% PVC; in the form of film and foam	Switzerland	[44]

fishes found in shallow oceanic habitats (epipelagic zone: 1–3 m depth) had higher MPs ingestion levels than species found in the deeper oceanic habitats (mesopelagic: >200 m depth). As a result, the epipelagic fish *(Rastrelliger kanagurta)* had higher MP exposure than the mesopelagic fish (*Istiophorus platypterus*) [21]. Similarly, Kumar et al. [22] also detected MPs size ranges of 0.5–1 mm in the gut and intestine of Rastrelliger kanagurta and Epinephelus merra fishes sampled from Thirespuram and Punnakayal landing sites of Tuticorin, India. The FTIR analysis has shown that 30% of the fishes samples contained polyethylene and polypropylene MPs particulates. The recognized MPs were existed in microfibers (80%) with red, black, and translucent colors, and fragments (20%) [22].

Another field study revealed the existence of MPs (>20-μm) in guts and gills of thirteen species of coastal fishes at Hangzhou Bay and Yangtze Estuary, China [23]. As depicted in Table 1, the result showed MPs concentration at an estimated percentage of (22–100%) in the guts and (22–89%) gills of fish collected samples. The mean abundance of MPs in the gut and gills of the fish samples varied from 0.3–5.3 items/individual (i.e., 0.1–8.8 items/g in the gut) to 0.3–2.6 items/individual (i.e., 0.1 to 5.2 items/g in the gill), respectively. But, MPs were not detected in the liver or muscle tissue of *Lateolabrax maculate.* Further to the occurrence of MPs, the μ-FTIR observation identified 10 polymer types and the dominant polymer was polyester, followed by polypropylene and polyethylene. The μ-FTIR-based polymer identification also validated that the most common plastics found in the gut and gill as PE and PP from polyesters (PES). Meanwhile, the shape, as well as size patterns of MPs, are different in gut and gill while fibrous MPs with small size are lodged in gill. In addition to plastic polymers, the non-plastic (cotton) from the muscle of the fish was identified which can be ascribed to background contagion of the fish [23].

Another investigation on MPs marine biota also shows the existence and effects of polystyrene (PS) exposure on Scrobicularia plana clam tissues [24]. Upon examination of MPs, the species were exposed to 1 mg/L of 20 μm PS for 14 days, followed by seven days of depuration. The infrared spectroscopy detected the presence of MPs in clam tissues in which 1 mg/L of PS could be accumulated in the gills and digestive gland. The effects of these MPs further evaluated through a battery of biomarkers and the result showed that MPs caused damage in antioxidant capacity, DNA, neuron, and oxidative damage of Scrobicularia plana. Furthermore, PS inhibits the Anti-cholinesterase (AChE) activity in clam gills even after seven days of depuration [24]. Furthermore, the chronic exposure to PS causes genotoxicity and the detoxification of PS microplastics in tissues is inefficient indicating potential trophic transfer [24].

The possibility of the occurrence of MPs through ingestion has been assessed in the early larvae stages of wild fish in the western English Channel. The FTIR analysis revealed the ingestion of MPs by fish larvae and identified 2.9% of wild fish larvae had ingested MPs. Most of the detected ingested MPs were nylon, polyamide-polypropylene, and rayon with 66% fibrous and blue while 34% of them were fragments and 16% red in color. In addition, the ingested microfibers have a strong resemblance to polymer characteristics found in water samples [25]. Likewise, Desforges et al. [26] inspected plastic ingestion by species of zooplankton

(*Neocalanus cristatus and Euphausia pacifia*) in the Northeast Pacific Ocean near British Columbia, Canada. The result proves the MPs encounter rates by ingestion were estimated to 1 particle/every 34 copepods (or 0.026 ± 0.005 particles/individual zooplankton) and 1/every 17 euphausiids (or 0.058 ± 0.01 particles/zooplankton). The confirmation of MPs ingestion thru by marine zooplankton indicated that species at lower trophic levels of the marine food web are muddle up plastic for food, which arises potential risks to higher trophic level species. Accordingly, the intake of MP containing zooplankton can lead to estimated ingestion of 2–7 MPs particles/day by single juvenile salmon in coastal British Columbia, Canada [26].

MPs were also detected in filter feeder and deposit feeder living organisms along the French–Belgian–Dutch coastline [27]. The uptake of MPs by marine invertebrates (blue mussel Mytilus edulis and lugworm Arenicola marina) was studied under field conditions collected from six locations of the study area at different feeding strategies. Under laboratory investigation, the mussels (filter feeder) and the lugworms (deposit feeder) were exposed to 110 particles/mL seawater and 110 particles/g sediments of polystyrene microspheres. The laboratory result illustrates the presence of MPs (<1 mm) in organisms together in the field: on average 0.2 ± 0.3 MPs/g of M. edulis and 1.2 ± 2.8 particles/g of A. marina. However, the aforementioned concentration of polystyrene microspheres didn't show an adverse effect on the overall energy budget of the species. [27]. Similarly, the occurrence of MPs was studied on species of commercially grown bivalves (*Mytilus edulis and Crassostrea gigas*). The investigation confirmed the ingestion of MPs (<1 mm) and recovered from both species' soft tissues. The MPs appeared as red particles extracted from M. edulis tissue resembles that of the pigment haematite. Whereas the blue particles from *C. gigas* resemble the widely deployed phthalocyanine dyes. For this reason, the Raman spectra obtained during the analysis could be originated from the pigments present in the particles, and not those from the plastic spectrometer. Furthermore, at the moment of human intake, *M. edulis* has an average of 0.36 ± 0.07 particles/g (wet weight of the organism), and C. *gigas* contains 0.47 ± 0.16 particles/g (wet weight of the organism). Consequently, the annual dietary exposure for European shellfish users is about 11,000 MPs per year. Hence, the occurrence of MPs in seafood could arises a risk to human health [28].

Moreover, MPs (<5 mm) uptake was noted in *Galeus melastomus*, the blackmouth catshark, nearby the Balearic Islands. In this regard, MPs abundance for 125 catshark samples were evaluated, and results shown 17% of the specimens had ingested at a mean value of 0.34 ± 0.07 MPs/individual. The identity and percentage of MPs polymers identified were Cellophane (33%), Polyacrylonitrile (4.5%), Polyethylene (4.5%), Polyethylene terephthalate (PET) (4.5%) Poly Ethyl Acrylate (27%) Polyacrylate (1.5%), Polyamide (PA) (12%), Polypropylene (PP) (3.0%), Alkyd (1.5%). In addition, higher magnitudes of filament type (86%), fragment (12%), and film (1.5%) MPs were recognized with black, blue, red, transparent, and white colors in *Galeus melastomus*. Furthermore, a percentage range of 0.86–39% stomach fullness index plus regression analysis indicated that fuller stomachs contained more MPs. The outcomes in the exploration reproduce the availability, quantity, and composition of MPs litter ingested by marine species in seafloor habitats [29].

Hoarau et al. [30] further studied the key problem arisen from marine plastic rubbishes triggered by anthropogenic letting on *Caretta caretta*, loggerhead sea turtles in the South-West Indian Ocean. According to the results, plastic debris was found with a mean abundance of 41 ± 7.2 particles per turtle. Among the turtles who ingested anthropogenic debris, 38 (51.4%) of plastic fragments existed in either gut or feces. In addition, fragments of plastic (96.2%) appeared with the mean percentages of 80 ± 3.69% hard plastics per turtle that ingested debris. Further, the incidence of additional caps and soft plastics categories was also observed accounted for a mean of 53.1% (11.29 ± 2.65 mean percent per turtle) and 71.9% (2.94 ± 0.79 mean percent per turtle), respectively. On the other hand, the ingested plastic debris was hard white and perfect plastics constituted together over semi of plastic debris content. The hard and blue plastics embodied 12.5 ± 2.59% and 9.61 ± 2.66% plastics ingested per turtle, respectively. But the physical appearance of ingested debris and biometric features of loggerheads have no substantial relations. Generally, the study shows a significant rate of the occurrence of anthropogenic debris due to ingestion by aquatic biota [30].

An assessment of marine feeder exposure has shown the occurrence of MPs in the intestines of a baleen whale (*Megaptera novaeangliae*) on a sandbank in Johanna, Netherlands. The FTIR analysis of the gastrointestinal tract described the existence of various polymer types (PE, PP, PVC, PET, and nylon) with variable particle sizes of 1 mm to 17 cm. This variety in polymer identity and particle size is inferred as a demonstration of the uneven features of marine plastic and the indiscriminate way of ingestion by *M. novaeangliae* [31]. In the other study by [32], an abundance of ingested debris by seals has been stated as a likely indicator of marine litter in the European Marine Strategy Framework Directive (MSFD). Meanwhile, the occurrence of plastic debris in stomachs, intestines, and scats (fecal) of harbor seals samples were analyzed in the Netherlands. Results proved the prevalence of plastic in stomachs (11%), in intestines (1%), and in scats (0%) and plastic debris mostly affects younger animals, up to 3 years of age [32].

Although MPs are a recognized pollutant in the marine aquatic system, low attention has been given to the freshwater ecosystem despite their greater vicinity to potential plastic sources. A recent study by Windsor et al. [33] quantified the occurrence of MPs in river organisms of macroinvertebrates (Baetidae, Heptageniidae, and Hydropsychidae) from five United Kingdom wastewater treatment works (WwTWs) across South-Wales river catchments. MPs were detected in 50% of macroinvertebrate at abundance up to 0.14 mg MP/tissue. But, MP concentration in macroinvertebrates slightly increased at more total runoff effluent discharges and declined with increasing river discharge [33]. Similarly, Su et al. [34] analyzed MPs uptake in a noxious fish species (*Gambusia holbrooki*) from nine wetland areas of Greater Melbourne area, Australia, size, weight, and gender of fish were also characterized in the study. As per their scrutiny, MPs were discovered in the head and other parts of fish with uptake rate direct proportion to size, and weight of fish. Furthermore, MPs were found in 19.4% of the sample fish with a quantity of 0.6 items/individual, and 7.2% of MPs found in the gills of fish with abundance estimated to 0.1 items/individual.

Female fish exhibited an affinity to ingest more MPs than male fish. The mainstream polymers constituted 25.7% PS, 10.1% rayon, 7.3% polyamide, and 5.5% PP. Particularly, polyester and rayon with fibers shape appeared greater than 70% of the wetland area, the other bodies of fish have more MPs than heads. Whereas the MPs identified in heads were lesser in size than those contained in another body of fishes [34].

In another freshwater system, the occurrence and distribution of MPs in water, sediments, and wild fish were investigated in the largest freshwater Poyang Lake, China [35]. The concentration of MPs <0.5 mm found 5–34 items/L in surface waters, 54–506 items/kg in sediments, and 0–18 items/individual in wild crucians (*Carassius auratus*). The majority of the identified polymer types were found PP and PE with predominant characteristics of fibrous MPs. These results confirmed the widespread occurrence of MPs in water, sediment, and biota of the Poyang Lake which will provide an insight for MPs contamination in inland freshwater systems [35]. Similarly, MPs pollution source and distribution were studied in the largest inland Qinghai Lake, China [36]. In the study area, MPs were discovered with abundance ranges 0.05×10^5–7.58×10^5 items/km^2 in the lake, 0.03×10^5–0.31×10^5 items/km^2 in the rivers, 50–1292 items/m^2 in the lakeshore sediment, and 2–15 items per individual in the fish samples. PE and PP were detected as a predominant polymer in Qinghai Lake. Moreover, the majority of MPs size were small MPs (0.1–0.5 mm) in the lake while large MPs (1–5 mm) are abundant in the river samples. Additionally, MPs were mostly sheet and fiber characters in the lake and river water unlike in the lakeshore sediment samples [36]. Bessa et al. [37] evaluated the occurrence of MPs in three commercial fish species (*Dicentrarchus labrax, Diplodus Vulgaris, and Platichthys flesus*) in Mondego estuary, Portugal [37]. In light of the evaluation, an aggregate of 38% fish species had ingested MPs with an average of 1.67 ± 0.27 particles per fish, the FTIR result also identified polyester, polypropylene, and rayon as the principal polymers found in commercial fish species. In terms of polymer shape, 96% of MPs found in the fishes were fibers followed by fragments (4%) and occurred in diverse size fractions with varying quantity. Whereas, ingested MPs color distribution was uniform throughout all fish species being mostly blue color (47%), transparent (30%), and black (11%). On the other hand, extra colors comprising red, yellow, and green were less frequently observed in the studied fish species [37].

Further to the ingestion of MPs, rivers are involved as major pathways of MPs transport to marine and lake ecosystems. For this evidences, McNeish et al. [38] studied MPs pollution in river food chains in connections to species feeding behavior from three major tributaries of Lake Michigan, USA. Accordingly, the MPs <1.5 mm in fish with a concentration of 10 ($\pm$ 2.3) to 13 ($\pm$ 1.6) particles per fish were reported. Fibers comprised about 97–100% of all MPs found in the water and fish. Fragments comprised the remaining 0–3% of MPs collected from water from the major tributaries of the lake and 0–3% in the fish. In contrast, fiber color patterns were similar in fish across sites, with clear and blue fibers predominant MPs size <1.5 mm size found across water samples and fish. This indicated that MPs litter is obvious in river food webs from point-source in addition to diffuse sources into aquatic ecosystems [38].

Similarly, the occurrence of anthropogenic particles such as MPs pollution through ingestion was examined in the liver and muscle fish (*Squalius cephalus*) taken from the Marne and Seine Rivers, France. The investigation showed 25% of sampled fish individuals had ingested 2–2.93 mm particle having a mean concentration of 0.36 particles/g (wet weight of fish). In addition, 5% of sampled livers contained greater than one MP particle. However, no MP particles were found in the muscle tissue. The dominant types of plastic polymers found in the stomach of the studied fish were polyethylene terephthalate (PET), polypropylene (PP), polyacrylonitrile (PAN), and polyethylene-co-vinyl acetate (PEVA). The majority of MP particles found in the stomach were fibers polymer. This points out that fish could have more exposure to fiber type of MPs contamination in the river [39].

Silva-Cavalcanti et al. [40] evaluated the ingestion of MPs by a freshwater fish (*Hoplosternum littorale*) consumed by humans in semi-desert of South America, Brazil. Primarily, the concentration, types of plastic debris, and other foodstuff were assessed in the gut of the fish. Finally, 83% of the sampled fish had plastic fragments (<5 mm) in their gut. In addition, 89% of the recovered plastic debris from the fish gut were fibers. According to the observation, 3.6 particles per fish of ingested MPs were found at the urbanized segments of the river. These suggest that freshwater biotas are susceptible to MPs litter due to rapid urbanization. The ingestion of MPs was negatively interrelated with the variety of other food items in the gut of individual fish [40]. Similarly, the incidence of MPs was studied in the fish gut of coastal freshwater in the Rio de la Plata estuary, Argentina. MPs size 0.06–4.7 mm was verified in all fish (100%). Moreover, 83% of collected fish contained plastic debris inside the guts at average number 18.5 $\pm$ 18.9 particles/fish and 0.7 $\pm$ 1.7 'others' MPs/fish. Furthermore, 96% of fibers and 4% of 'others' MPs were found in 1679 pieces sample [41].

Likewise, the ingestion of MPs by freshwater benthic macroinvertebrates (*Tubifex tubifex*) was demonstrated in River Irwell, UK. The result demonstrated the ingestion of the mean value of 129 $\pm$ 65 particles/g tissue of the invertebrate. The findings proved that 87% of the ingested MP by *Tubifex* were microfibers (55–4100 μm), while the rest 13% were fragmented (50–4500 μm) containing polyester and acrylic fibers. The observation also shows that Tubifex worms hold on MPs for longer than other particulate constituents of the consumed sediment matrix. In the end, MPs ingestion by *Tubifex* causes a major threat for trophic transfer as well as biomagnification of MPs through the food chain [42].

Later, the bioavailability of MPs was studied in Asian clams, water, and sediments of the Middle-Lower Yangtze River Basin, using *Corbicula flumine* species as a bioindicator [43]. The result has shown the presence of MPs at a concentration of 0.3–4.9 particles/g which equivalently 0.4–5.0 particles/individual in clams, 0.5–3.1 particles/L in water, and 15–160 particles/kg in sediment. Moreover, the μ-FTIR confirmed that polyester (33%) was the dominant polymer type in the clams, followed by polypropylene (19%) and polyethylene (9%). So far, blue and transparent items consisted greater than 30% of all particles, and the rest 70% of the MPs found in clams were 'other colors'. Microfibers (0.25–1 mm) were the major MPs types that accounted for 60–100% found in clams throughout sampling locations. In addition, there was a great similarity in the abundance, size distribution, and color patterns of MPs in clams those in sediment than in water [43].

Zhang et al. [44] examined MPs' occurrence in biota, distribution, and physical characteristics in the backwater area of Xiangxi River, China. Most importantly, polyethylene and nylon (0.3–1.8 mm) MPs were detected in the digestion tracts of 26% of fish samples. In addition, MPs were detected in both surface water and sediment with concentrations ranging from 0.55 × 105 to 342 × 105 particles/km^2 and 80 to 864 particles/m^2, respectively. Specifically, PE, PP, and PS polymers were identified in surface water, while PE, PP, PET, and pigments were observed in sediment [44]. Additionally, Faura et al. [45] assessed plastic abundance, potential ingestion by birds and fish, and the associated pollutants in Lakes Geneva, Switzerland. Evidence shows the abundance of MPs (<5 mm) in 4.3 ± 2.6 particles per bird found in all samples. The result suggests birds and fish are susceptible to MPs ingestion [45].

4 Interaction and Uptake Mechanism of Microplastics by Aquatic Biota

MPs finally entered into water bodies by surface runoff, discharge from wastewater plants, and domestic/industrial drainage systems. MPs interaction with the surrounding environment and aquatic biota is dynamic. After entering the aquatic systems, MPs distributed into numerous environmental matrices such as surface water, and benthic sediment causes bioavailability to the aquatic biota. Besides, MPs are continually changing their fate and bioavailability in the aquatic system. The vast majority of MPs in the oceans are supposed to originate from the weathering of larger plastic debris [46], through mechanical and biological degradation. The weathering of the large plastic debris is driven mainly by UV-radiation-induced photooxidation, releasing low-molecular-weight polymer fragments resulting in fragmentation to a range of smaller [47]. Then, these weathered MPs have contaminated the aquatic biota inhabitants and interact with biota in different levels through ingestion and tropical transfer as illustrated in Fig. 1 mainly through ingestion of plastic litter.

Ingestion is the most likely integration and uptake mechanism of MPs by marine and freshwater biota [7]. Ingestion of plastics can be direct and indirect [19]. Laboratory tests have shown that amphipods and sea cucumber, barnacles mussels, and lugworms can ingest MPs as the particles are in the size range of their prey [48]. Therefore, small plastic fragments have existed to invertebrates at the base of the food shown in Fig. 1. Hence, direct ingestion is inherent in food consumption resulting from unintentional intake through non-preset criteria of ingestion (by filter feeders), or active selection to eat plastics instead of food. Small-size MPs give them the possibility to be ingested by a wide range of biota in benthic and pelagic ecosystems. Particles can also accumulate in sediment [16] suggesting that these would be available to many benthic communities. In some cases, the aquatic biota feeding method does not permit a distinction between prey and anthropogenic items [49]. Secondly,

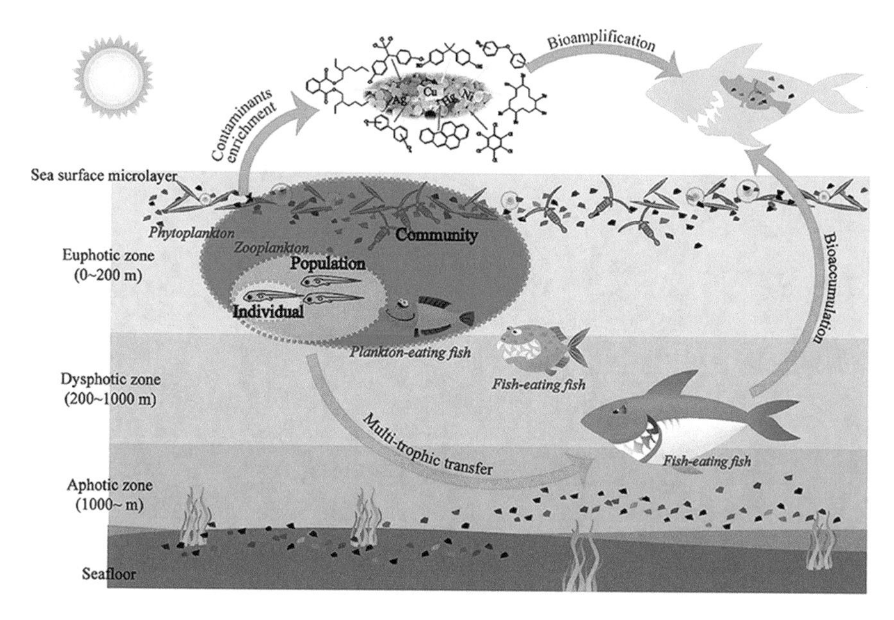

Fig. 1 The overall uptake mechanism, ingestion and tropical transfer of microplastics by aquatic biota [57] adapted with permission of Elsevier, 2020

aquatic biota might feed directly on MPs, mix up with their prey, or intentionally feed on MPs in place of food [50]. As indicated in Fig. 1, MPs in the size of phytoplankton along with persistent organic pollutants could be ingested and facilitate bioaccumulation in an aquatic organism. Consequently, exhibits biomagnification toward humans through the food chain as depicted in Fig. 2.

The other integration and uptake mechanism of MP by an aquatic organism has been through the entanglement of marine fauna, ranging from zooplankton to cetaceans, seabirds, and marine reptiles [5]. Entanglement is indirect plastic ingestion through trophic transfer as the result of ingesting contaminated prey by predators [51]. As showed in Fig. 1, the plankton-eating fish directly ingested zooplankton and which in turn was eaten by the higher trophic level. In addition, field observation showed the occurrence of MPs in the scat of fur seals (*Arctocephalus spp.*) proposed that MPs had firstly ingested by the fur seals' prey to the plankton feeding Mycophiids [52]. In feeding experiments of [53] identified MPs in the gut and hemolymph of the shore crab (*Carcinus maenas*), which had previously been ingested by blue mussels (*Mytilus edulis*). Therefore, MPs found in the scat of fur seals (*Arctocephalus spp.)* were assumed to have been eaten by lantern fish, which were in turn eaten by the seals [53]. Similarly, Nephrops-fed fish, which had been planted with MPs strands of polypropylene rope were found to eat but not to excrete the strands, again implying potential trophic transfer [54]. Another study has demonstrated the potential polystyrene microspheres (10-μm) transference from different Baltic Sea mesozooplankton taxa at much lower concentrations. Several studies detected the prospective

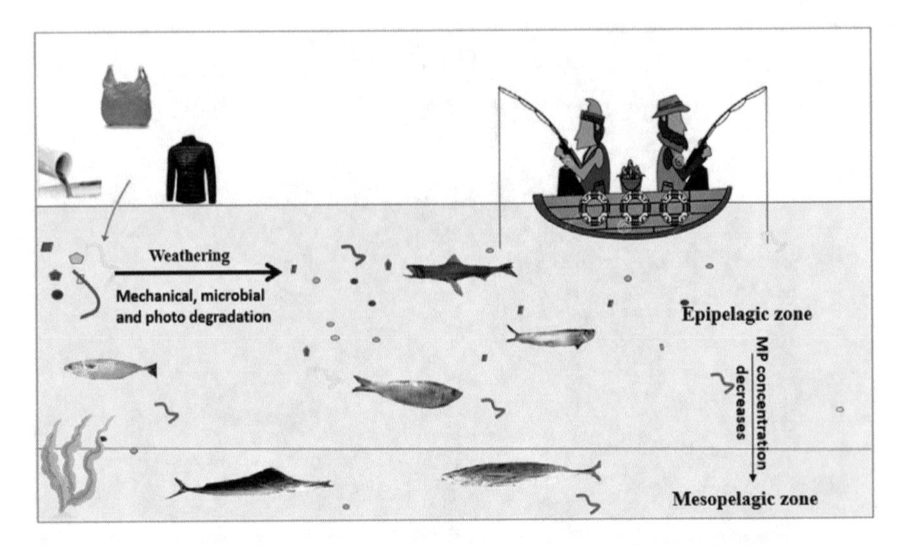

Fig. 2 Contamination of microplastics to seawater and fish from different habitats and level of human exposure to microplastics [21] adapted with permission of Elsevier, 2020

of MPs transfer via planktonic organisms from one trophic level to a higher level as shown in Fig. 1 [55]. Later, the MPs were also discovered in human consumed seafood such as cod, whiting, haddock, bivalves, and brown shrimp, which in turn, raises concerns for trophic transfer to human exposure [56]. Particularly, the incident of weathered plastic fragments in the most prevalent fish species raises human health issues due to the consumption of contaminated fish as indicated in Fig. 2, which could probably surge the precarious chemicals in the human body.

5 Ecotoxicological Effects of Microplastics on Aquatic Biota

MPs once ingested, can distract aquatic life in several ways and chronic exposure to MPs is rarely lethal [58]. The toxic effects of MPs on individual animals is mostly demonstrated by physical damage. These consecutively retards the growth of organisms, and even causing death [12, 59]. Research investigations evidently disclosed that marine biota exposed to MPs at a significant level desperately changes the function and variety of the aquatic organisms. Consequently, the effects of MPs have been detected at different trophic levels of the biological group. It is clear that few investigations on the toxicity of MPs to aquatic biotas have been conducted and exposure to MP at consecutive stages of the biological organization was summarized in Table 2. Ultimately, the MPs have widespread throughout the aquatic ecosystem affecting the

Table 2 Toxicological effects of MPs in marine and freshwater biota species

Water body	Studied biota species	MPs type	MPs size	MPs concentration range	Exposure time	Reported biological effects	References
Marine environment	*Zebrafish (Danio rerio)*	PS	0.07–20 μm	20–2000 μg/L	3 weeks	MP accumulation in gills, liver, and gut results in surges in enzyme activities such as superoxide dismutase and catalase. Considerable changes in hepatic metabolites, penetration, and lipid droplets in hepatocytes	[59]
	Zebrafish (Danio rerio	PS	5-μm	50 μg/L and 500 μg/L)	3 week	Redness and oxidative stress in the gut. Change in the gut microbiome and tissue metabolic profiles, linked with oxidative strain, swelling, and lipid metabolism, and gut damage	[78]
	Adult male zebrafish	PS	0.5 & 50 μm	1.456×10^{10} particles/L for 0.5 μm and 1.456×10^{4} particles/L for 50 μm).	14 days	A paramount alteration in the fertility and variety of microbiota in the gut, changes in gene level, increased mRNA levels of microbiota imbalance, and redness in the gut	[61]
	Shore crab Carcinus maenas	PS	8 μm	–	1 h	Decline in Na^{+} ion and rise in Ca^{2+}·ion in hemolymph	[62]

(continued)

Table 2 (continued)

Water body	Studied biota species	MPs type	MPs size	MPs concentration range	Exposure time	Reported biological effects	References
	Brachionus koreanus	PS	0.05-, 0.5-, and 6-μm	0.1, 1, 10, and 20 μg/mL	12 days	Growth, fertility, and lifespan reduction, also brings longer reproduction time. Increases the ROS levels	[63]
	Scrobicularia plana	PS	20 μm	1 mg/L	14 days	Leading impacts on antioxidant capacity cause DNA, neurotoxicity, and oxidative damage	[24]
	Daphnia magna, Daphnia pulex, Ceriodaphnia dubia	Spherical fluorescent PM SPM from polyethylene microspheres	PMP; 1–5 μm, SMP; 1–10 μm,	10^2 p/mL–10^5p/mL	21 day for PMP 7 day for Ceriodaphnia dubia	Adverse effect cumulative number of neonates, the total number of broods, and terminal body length of test animals	[64]
	Crustaceans (Amphibalanus amphitrite barnacle and of Artemia franciscana brine shrimp)	PS	0.1 μm	0.001–10 mg/L	24 and 48 h.	Neurotoxic effect and oxidative stress, low enzymatic activity, interrupted the detoxification of hydrogen peroxide in the injured organisms	[65]

(continued)

Table 2 (continued)

Water body	Studied biota species	MPs type	MPs size	MPs concentration range	Exposure time	Reported biological effects	References
	Silver barb (Barbodes gonionotus)	PVC	0.1–1000 μm	0.2–1.0 mg/L	96 h	The mean distal and intestine size of the fish enlarged by 73.4% and 29.1%, respectively Increased the activities of enzymes mainly trypsin and chymotrypsin	[67]
	African catfish (Clarias gariepinus)	PE	<60 μm	50–500 μg/L	96 h	Significantly decreased the levels of blood HDL. Increase the water-soluble protein to salt solution soluble protein ratio of fish. In addition, considerably decreases the dictation levels of tryptophan hydroxylase-2	[68]
	Jacopever (Sebastes schlegelii)	PS	15 μm	10^6 items/L	21 days	Significantly reduced swimming haste and wide-ranging movement of fish. Decreases weight gain rate by 65.4%, specific growth rate by 65.9%, and gross energy of fish by 9.5%	[69]

(continued)

Table 2 (continued)

Water body	Studied biota species	MPs type	MPs size	MPs concentration range	Exposure time	Reported biological effects	References
	Asian green mussel Perna viridis	PVC	1–50 μm	0 mg/l, 21.6 mg/l, 216 mg/l, 2160 mg/l)	91 days	Decreased the respiration rates, byssus production as well as mussel survival with increased PVC abundance	[72]
Freshwater	*zooplankton Ceriodaphnia dubia*	PS fibers and PE beads	100–400 μm	0.5 to 16 mg/L of PE beads and 0.125 to 4 mg/L of polyester fibers 62.5 to 2000 μg/L for PE beads and 31.25 to 1000 μg/L for polyester fibers	48 h, 8 day	The acute exposure induced 40% mortality in higher concentrations. Whereas, Chronic exposure reduced body size and the number of neonates. The more pronounced effects have been observed on effect PS fibers	[73]
	Hyalella azteca	PP; PE	10–27 μm	0–20,000 mg/mL	10 day and 42 day	The exposure for 10 days causes mortality while 42-day chronic exposure reduced growth and fertility in the range of low to intermediate concentrations exposure	[74]
	Hydra attenuata	PE	<400 μm	0.01–0.08 mg/L	30 min; 60 min	Significant reductions in feeding rate, morphological changes, and reproduction	[75]

(continued)

Table 2 (continued)

Water body	Studied biota species	MPs type	MPs size	MPs concentration range	Exposure time	Reported biological effects	References
	Chlorella and Scenedesmus	PS	20 nm	0.08–0.8 mg/mL	2 h	Adsorbed algae in PS beads causes CO_2 depletion	[76]
	Scenedesmus obliquus	PS	70 nm	44–1100 mg/L	72 h	Reduction in Algae growth and chlorophyll-a concentrations	[77]
	Caenorhabditis Elegans	PA PE PP PVC PS	70 μm 70 μm 70 μm 70-μm 0.1, 1.0, 5.0 μm	0.5–10.0 mg/L	2d	4.89–11.44% decreases in the mean body length, inhibition rates in embryo number, and brood size by 14.40–25.22% and 2.44–28.02%, respectively	[70]
	zebrafish Danio rerio	PA PE PP PVC PS	70 μm 70 μm 70 μm 70 μm 0.1, 1.0, 5.0 μm	0.001–10.0 mg/L	10d	Reduction of 27.1% in survival, causes 73.3–86.7% damage in the intestine of individuals	[70]

individuals, then the population, and finally disturbing the community structure and function as a whole. Consequently, MP pollution in the water ecosystem poses a risk to biota across the range of cells to the population level of biological organization [12].

5.1 *Toxicity Effect at Subcellular, Cellular, and Organ Level of Biota*

Several studies have intensively reported the individual-level effects of MP ingestion in adult organisms at the cellular and subcellular levels. For example, the ingestion and tissue accumulation of polystyrene MP (PS-MP), and its toxicity were detected in liver organs of [60]. The experimental investigation of PS-MP exposure for seven days showed the accumulation of 5 μm PS in the gill, liver, and gut, whereas 20 μm MP was found in fish gill and gut only. Further, the histopathological analysis indicated that the smallest sizes of 5 μm and 70 nm PS-MP caused redness and lipid buildup in the fish liver. Moreover, PS-MP also considerably increased superoxide dismutase (SOD) and catalase (CAT) activities in fish liver. In addition, exposure prompted modifications of metabolic profiles in fish liver and distressed the lipid and energy metabolism [60]. Similarly, impacts of ingested PS-MPs (5-μm beads, 50 μg/L, and 500 μg/L) on zebrafish were examined in the gut after 21 days of the exposure period. The result showed a noticeable adverse impact of PS-MPs ingestion by zebrafish on the adult growth rate, fecundity. The internal exposure of MPs to the circulatory system, tissue, and organs affect metabolic disorders and sublethal responses in organisms. Consequently, the effects result in endocrine disorders, oxidative stress, immune responses, and gene expression alteration [61]. Furthermore, the metabolism and growth of an adult male zebrafish were critically affected by PS-MPs. As shown in Table 2, the ingestion of the PS-MP (0.5 and 50 μm) in the guts of the adult zebrafish enlarged the volume of mucus and triggered inflammation after 14 days of ingestion. The high-throughput sequencing of the 16S rRNA gene V3–V4 region and the operational taxonomic unit analysis discovered an in-depth response at the phylum and genus levels, suggesting that the MP exposure within the gut changed the richness and variety of the microbiota, leading to a microbial imbalance [62]. Further to chronic exposure of MPs, Watts et al. [63] examined that acute aqueous exposure to PS-MP (8 μm) ingestion and inhalation by the shore crab *Carcinus maenas*. The result has shown that exposure to PS-MP has a little degree of effects on the oxygen consumption of the shore crab *Carcinus maenas*. However, a significant reduction in the hemolymph sodium ions and a rise in the calcium ions were observed in cellular metabolism [63].

In other studies, Jeong et al. [64] assessed accumulation and antagonistic impacts of ingestion of 0.05-, 0.5-, and 6-μm polystyrene in microbeads monogonont rotifer (*Brachionus koreanus*). The analysis of the introduction of all sizes of polystyrene microbeads caused impacts on the growth rate, fertility, life expectancy, and reproduction time. The rotifers exposed to the smallest particles (0.05-μm) were effectively egested and displayed more toxicity effects on growth to rotifers. Similar to the effects on growth, the rotifer fertility rate was affected by different sizes of microbeads. Similarly, the smallest particles (0.05-μm) microbeads exerted the most toxic effects on the fertility rate. Therefore, the findings have shown microbeads toxicity is size-dependent and lesser size microbeads were more toxic. Consequently, the 0.05-μm microbeads increase reactive oxygen species (ROS) levels in the monogonont rotifer *Brachionus koreanus* [64].

Likewise, the effect of PS-MPs (20-μm, and 1 mg/L) exposure on clam *Scrobicularia plana* were assessed for 14 days, followed by seven days of depuration. The results revealed that PS-MP induces effects on antioxidant capacity, DNA damage, and the nervous system [24]. In other investigations, reproductive toxicity of the chronic exposure of 1–5 μm MP and 1–10 μm MP was assessed on Cladoceran species (*Daphnia magna, Daphnia pulex, andCeriodaphnia dubia*). The results have shown that the fertility of species dropped due to the ingestion of both size ranges of MP. In particular, chronic exposure to concentration 10^2–10^5p/mL of MP impaired reproductive output of the number of neonates by influencing brood sizes. The study also disclosed that 1–5 μm MP particles are more toxic to the studied organisms compared to 1–10 μm MP [65].

Similarly, the toxic effects of 0.1 μm PS beads were investigated using oceanic *planktonic crustaceans*. The experimented *Crustaceans* species showed the accumulation of MP without causing mortality of larval stages of the species. However, swimming activity was significantly affected in crustaceans exposed to high MP concentrations (>1 mg/L) after 48 h. In addition, enzyme activities were meaningfully affected by all marine crustaceans due to the presence of MPs, indicating the introduction of neurotoxic effects and induction of oxidative stress in organisms [66, 67]. Correspondingly, the toxicity of the exposure to PVC fragments concentration of 0.2, 0.5, and 1.0 mg/L for 96 h was evaluated for fry fish. According to the whole body histological assessment and examination of the digestive enzymes trypsin and chymotrypsin, the activities of the enzymes increased considerably in fish exposed to higher concentrations of PVC (0.5 and 1.0 mg/L). But no tissue damage was apparent in other interior body parts or gills. Henceforth, The mean thickness of the fish distal and proximal intestine increased by 73.4% and 29.1%, respectively [68].

Another impact study of phenanthrene (Phe)- loaded LDPE fragments were conducted using juvenile African catfish (*Clarias gariepinus*). The virgin LDPE concentration of 500 mg/L and 100 mg/L were exposed to the fish for 96 h. The study showed the phenanthrene usages ominously increased the extent of tissue alteration in the liver while reduced the transcription levels of forkhead box and tryptophan hydroxylase in the brain of *C. gariepinus*. Consequently, the exposure to either concentration of the stated MPs increased the degree of tissue change in the liver and plasma albumin while reduced the transcription extent of tryptophan hydroxylase. Moreover, due to the widespread of MPs and other associated pollutants in marine environments, virgin LDPE fragments can able to trigger toxicity and modulate the harmful effects of Phe in *C. gariepinus* [69].

In the same manner, Yin et al. [70] also were examined the effects of PS-MPs (1×10^6 microspheres/L) on the intrinsic behavior, energy reserve, and nutritional composition of juvenile jacopever (*Sebastes schlegelii*) for 3 weeks. The fish were treated with PS and showed inferior sympathy to the added food in the tank, and greatly declining feeding activity of the fish. Remarkably, the MPs treated-fish noticeably reduced swimming haste and variety of movement, proving that PS-MPs could harm hunting behavior. Furthermore, PS-MPs accumulated in the gills and intestine, triggering major histopathological alterations in the gallbladder and liver [70].

Lie et al. [71] explored the toxic effects of five types of MPs: PA, PE, PP, PVC, and PS particles on *Caenorhabditis Elegans and zebrafish Danio rerio.* Results showed an exposure concentration of 0.001–10.0 mg/L specified MPs for 10 d has no or low lethality in D. rerio. However, such MPs with ~70-μm size instigated cracking of villi as well as splitting of enterocytes leading to internal organ damage of *D. rerio*. In addition, exposure to 5.0 mg/L above-mentioned MPs for 2 d remarkably inhibited survival rates, body length, and reproduction of *C. elegans*. Moreover, exposure to MPs decreased calcium levels and increased the appearance of the glutathione S-transferase enzyme in the intestine in *C. elegans*. Among MPs (0.1, 1.0, and 5.0-μm) sizes of PS, the 1.0-μm particles produced the maximum mortality, the maximum buildup of PS in nematodes, the lowest Ca^{2+} level, and the maximum appearance of glutathione S-transferase in nematodes. This shows the effects and toxicity of MPs closely dependent on their size [71].

5.2 *Toxicity Effects on the Behavioral Pattern*

The disorder of behavior, fertility, and survival of aquatic biota is considered as the final toxicological effect under stress from anthropogenic adaptations to the environment [72]. To point out severe toxicology impacts of MPs, Wong and Candolin [72] investigate a potential exposure pathway of (PVC) particles (1–50-μm) to Asian green mussel *Perna Viridis*. Indeed, the mussel was exposed to concentrations ranges 0 mg/l–2160 mg/l in ten folds intervals of PVC for 2-hour-time-periods per day. Later on, after 91 days of exposure to PVC, mussel survival declined with increasing PVC concentration. In a similar investigation, PVC caused the mortality of the Asian green mussel *Perna Viridis* due to remarkable energy reserve depletion [73].

Similarly, Ziajahromi et al. [74] studied the acute and chronic toxicity of PS fibers and PE beads on freshwater zooplankton *Ceriodaphnia dubia* for 48 h and 8 days, respectively. Meanwhile, acute exposure to fibers and beads indicated a negative impact on the survival of species in a dose-dependent manner. The binary combination of beads and fibers caused antagonistic effects than additive effects. But, the acute exposure of *C. dubia* to PE beads and polyester fibers induced 40% mortality with increasing concentrations of the polymers. On the other hand, chronic exposure to fibers indicated more adverse effects than PE beads. In the long run, the exposure to fibers MPs poses a severe risk to *C. dubia*, with reduced fertility rate at environmentally relevant concentration. As a result, the chronic exposure of *C. dubia* to MP fibers on energy loss and physical damage, leading to reduced growth and reproduction. Moreover, chronic exposure of both PE beads and PS fibers leads to a reduction in body size and the overall quantity of neonates. In addition, abnormal swimming behavior was only observed in *C. dubia* exposed to polyester fibers, with their movement often inhibited as a result of entanglement in twisted fibers. Particularly, reduced reproduction and growth are associated with the inability to tolerate fibers in the environment, loss of energy to physical contact with fibers during chronic exposure to fibers [74].

Further study was also carried out to examine the toxicity of MPs ingestion by amphipod, exposed to polyethylene (PE) particle and polypropylene (PP) microfibers of 0–20,000 d/mL for 10 and 42 days of exposure. The 10-days toxicity of PP MPs fibers was significant than PE MPs particles. In addition, chronic exposure (42-days) of PE particle and PP fibers in the same concentration had shown decreased growth and reproduction at the low and intermediate exposure concentrations. But, MPs fibers had higher toxicity than MPs particles due to the longer residence time accumulation of fibers in the gut. Consequently, the variance in residence time could affect the capability to process food, ensuing in an energetic effect revealed in sublethal endpoints [75]. Likewise, the effect of PE MPs ingestion on the freshwater cnidarian, *Hydra attenuata* at different concentrations (0.01, 0.02, 0.04, and 0.08 g/mL) were evaluated. The results of the study [76] demonstrated that *Hydra attenuata* ingest exposed MP. Accordingly, a significant decline in feeding rates was observed after half an hour and 1 h MPs exposures. Therefore, the exposure to the MPs produced remarkable alterations to the morphology of *Hydra attenuata*, however, these alterations were non-lethal [76]. In other investigations, physical adsorption of PS was studied on freshwater primary producers such as *Chlorella* sp. and *Scenedesmus* sp. The adsorption was found diligently to favor positive charges over negatively charged plastic beads as a result of the electrostatic attraction among the beads and the cellulose component of the living systems. As a result, photosynthesis Nano-MPs adsorbed on *Chlorella and Scenedesmus* from physical blockages on light, air, and CO_2 depletion. Furthermore, adsorbed PS beads on algae hinder photosynthesis and promote the generation of reactive oxygen species [77]. Correspondingly, the toxicity of Nano-polystyrene (nano-PS) was evaluated for the green alga *Scenedesmus obliquus* and the zooplankton, *Daphnia magna.* Evidence has shown that nano-PS reduced growth and chlorophyll concentrations in the algae [78]. Similarly, nano-PS triggered a reduction in body size and severe modifications in the reproduction systems of *Daphnia* sp. The quantities and body size of neonates rose to 68% of the entities. Furthermore, the survival of *Daphnia* was reduced with an average of 27% mortality and 11% reduction in *Daphnia* body size upon exposure to MP concentration in a range of 0.22 to 103 mg nano-PS/L. The presence of such a concentration of nano-PS leads to life-history changes in aquatic organisms such as algae and Daphnia [78].

6 Microplastics Associated Contaminants Effects on Biotas

Plastics can constitute a variety of additives. Phthalates are among the extensively utilized plastic additives and are often allied with PVC [80]. The interaction of additives or contaminants between MPs and the nearby water causes adsorption of contaminants on plastics due to the intrinsic hydrophobic nature of the MPs [80–82]. Thus, the adsorption of pollutants such as organic or inorganic in the water is one of the most common behaviors of MPs due to its large surface area and strong hydrophobicity of MPs particles [83]. Moreover, earlier studies have been attested to the adsorptive binding of MPs on heavy metals, PAHs, PBDEs, PCBs, and DDT [6, 84, 85]. Therefore, MPs are allowing them to accumulate the aforementioned organic and inorganic pollutants [6, 80, 86]. In addition, organic contaminants can generally adsorb to the non-crystalline regions of plastic polymers with tiny additives that tend to leave the plastics fastest [80]. For example, a study by Bakir et al. [87] showed that phenanthrene and 4,4-DDT reached sorption equilibrium on plastics relatively quickly within 24 h. Furthermore, these persistent organic pollutants (POPs) shown to adsorb onto MPs at concentrations that are several magnitudes higher than in the surrounding water surges the revelation of aquatic organisms to MPs associated contaminants [80, 81, 88, 89]. Likewise, inorganic substances such as metals can adhere to plastic particles, which can accumulate at absorptions comparable to, or more than, those in the sediments or water [90].

Besides the adoption of pollutants, MPs can constantly release additives such as plasticizers, stabilizers, pigments, fillers, and flame retardants in the water ecosystem. Some of the additives are shown to be toxic, carcinogenic, or endocrine disruptors [91]. For example, Fries et al. [92] tested various plastic additives in MPs and found the occurrence of phthalates whereas Wagner and Oehlmann verified plastic released endocrine-disrupting chemicals (EDC) [93, 94]. These additive chemicals may then migrate with MP via the food chain to higher concentration, such as bisphenol A (BPA) and phthalates, can also potentially disturb the endocrine systems of aquatic organisms lead to impact mobility, fecundity, and growth. Phthalates and BPA are well-known endocrine disruptors in fish, and invertebrates, and were shown to cause whole body and molecular effects at concentrations in the ng/L to mg/L range [6].

Studies have been conducted on combined toxicities effects MPs in with a carboxyl group (PS-COOH) and PS combined with heavy metal nickel (Ni) on *Daphnia magna in* recent decades. The test result presented that Ni can be adsorbed on PS-COOH and PS has toxicity in combination with either of the PS-COOH or PS. As illustrated in Table 3, the acute toxicity test PS exhibited a negligible effect on Ni toxicity, whereas PS-COOH had an additive interaction effect with Ni. Furthermore, the immobilization of *D. magna* exposed to Ni coupled with PS-COOH was higher than that of *D. magna* exposed to Ni coupled with PS. But, the effects of MPs and pollutants may vary depending on the specific properties of the pollutant and MPs functional groups [95]. Other studies of MPs and associated contaminants effect on biota have been investigated on juvenile fish to MPs concentration of 0.26 and

Table 3 Microplastics and associated contaminants' effects on aquatic biotas. Modified from [57]

Type of chemical	Microplastic	Associated pollutants	Species name	MPs toxic effect only	Joint MPs and associated pollutant effect	References
Inorganic	PS/ PS-COOH	Ni	*Daphnia Magna*	Irregularities in immobilization and modification in the morphology of the species	The acute toxicity of PS showed an insignificant or less than they would be expected toxic effect on Ni toxicity, while PS-COOH had a synergetic effect with Ni than the individual any of them	[94]
	Fluorescence red polymer microspheres	Hg	*European seabass Dicentrarchus labrax*	The metal found in the brain as well as in the muscles triggering neurotoxicity and oxidative stress in the species. Finally indeed with damage, and alterations in enzyme activities	Pre-adsorption of Hg on the microplastic results in a gradual buildup of the pollutant Hg in tissues of the species	[95]
	PS	Cu	*zebrafish Danio rerio*	The MPs had gradually gathered on tissue and produced toxicity to guts and liver of the fish	MPs and NOM intensified the increase in quantity and poisonous of Cu. Increase the levels of $CH_2(CHO)_2$ (malonaldehyde) and metal-binding proteins involved in metabolism copper in the body tissue, and decreases the enzyme (superoxide dismutase) that helps the breakdown of harmful oxygen molecules	[78]

(continued)

Table 3 (continued)

Type of chemical	Microplastic	Associated pollutants	Species name	MPs toxic effect only	Joint MPs and associated pollutant effect	References
	PE	Ag	*rainbow trout Oncorhynchus mykiss*	MPs themselves do not affect but means for the conveyance of the pollutant; as a result, Ag was accumulated in between lower parts of alimentary canals of the mucus and muscle layer, mucosal epithelium, and serosa	Ag was accumulated in intestinal compartments and doesn't have toxic effects alone. The effect of the MP vector hypothetically brings together likely toxin forms into the intestine	[96]
	PE	PCBs	*Lugworm Arenicola marina (L.)*	Causes gradual accumulation of PCBs on the organism and impact feeding activity	MP brings a bit to a gradual accumulation of PCBs	[97]
	PVC	Chiral antidepressant venlafaxine and its metabolite Odesmethylvenlafaxine (pharmaceuticals)	*Loach Misgurnus anguillicaudatus*	Assisted the transferal and increments in contaminants to the liver through time	Delay the pollutants chemical reaction that occurs in the organism	[98]

(continued)

Table 3 (continued)

Type of chemical	Microplastic	Associated pollutants	Species name	MPs toxic effect only	Joint MPs and associated pollutant effect	References
Organic	PE microbeads	TCS HOC	*Marine copepod Acartia tonsa (Dana)*	Prompted accumulation of contaminants, metabolic reactions, and death	MP increase TCS-involved toxicity owing to the adsorption capacity	[99]
	PE, PS, PVC PVC800	TCS	*Microalgae Skeletonema costatum*	Caused physical damage while the triclosan causes growth inhibition on microalgae	TCS and MP had antagonistic toxicity effect and is increased with the greater adsorption capacity of TCS	[100]
	Red fluorescent polymer microspheres	Procainamide and doxycycline (pharmaceuticals)	*Marine microalga Tetraselmis chuii*	Caused a decrease in growth rate and the chlorophyll concentration	The interaction of pollutant with MP augmented the harmful poisonous effect	[101]
	Fluorescent microplastic particles	POPs polycyclic aromatic hydrocarbon benzo[a]pyrene (BaP)	*Artemia nauplii and zebrafish*	Causes a release of an adsorbed substance in the intestine and relocated to the intestinal epithelium as well as liver	Act as a carrier for contaminant conveyance at diverse trophic levels	[102]
	PS plastic particles	^{14}C-phenanthrene	*Daphnia Magna*	Inhibited the dissipation and alteration of phenanthrene in the medium	The toxicity increased due to the higher adsorption capacity of phenanthrene on PS	[103]

(continued)

Table 3 (continued)

Type of chemical	Microplastic	Associated pollutants	Species name	MPs toxic effect only	Joint MPs and associated pollutant effect	References
	PE PS	PAHs	*mussel* *Mytilus galloprovincialis*	Accumulation in hemolymph, gills, and digestive issues Changes in immune reaction, lysosomal section, peroxisomal production, antioxidant, neurotoxic effects, the beginning of genotoxicity	PAHs adsorption uplifts availability and toxicity mechanisms of the chemicals	[107]
	PE	PAHs/PCBs/PBDEs	*Japanese medaka* *Oryzias latipes*	Root for change in gene expression	Causes malfunctioning of the endocrine system of aquatic organism	[104]
	PVC	nonylphenol, triclosan, phenanthrene, and PBDE-47	*lugworms*	Causes greater than 30% and more vulnerable to oxidative stress	Reduced disease-causing bacteria removal capacity of coelomocytes by >60% Reduced the ability of worms to cause sediments and instigated death by >55%	[105]
	PS	PCBs	lugworms	reduction in feeding activity	no effects on survival, but PCBs within tissues raised by 1.1–1.5 factors	[106]

0.69 mg/L and mercury concentrations of 0.010 and 0.016 mg/L. Mercury accumulated in the brain and muscle, with significant interaction with MPs. In addition, all the mixtures exposure to MP and mercury through the water caused notable inhibition of brain acetylcholinesterase (AChE) activity by 64–76%, and substantial rise of lipid oxidation (LPO) levels to 2.9–3.4 folds in the brain and 2.2–2.9 folds in muscle. The investigation proved that MPs and mercury caused neurotoxicity through AChE inhibition, increased (LPO) in brain, muscle, and altered the activities of enzymes of lactate dehydrogenase (LDH) and isocitrate dehydrogenase (IDH). Overall, the study [96] has shown that MPs affect the bioaccumulation of mercury by *D. labrax* juveniles; the combination of MPs and mercury in parts per billion (ppb) concentrations can be a source for neurotoxicity, oxidative stress and damage, and alterations of enzymes activities in juveniles species [96]. Similarly, Qiao, Lu, et al. [79] evaluated the combined toxicity of MPs and organic matter with copper (Cu) in zebrafish (Danio rerio). The study of [79] revealed small-size MPs absorb extra Cu than large-size MPs. Further, the occurrence of organic matter supported Cu adsorption on MPs in the 6–8 pH environments. Furthermore, it has been demonstrated that the mixture of MPs and organic matter enhanced Cu accumulation in the liver and gut. Similarly, the outcomes of the study presented that MPs and NOM exacerbate Cu-toxicity in the liver and gut. Consequently, the fish demonstrated increased levels of malonaldehyde (MDA) and metallothionein (MT) and decreased levels of superoxide dismutase (SOD). Additionally, the transcriptomic analysis suggested that the toxicity exacerbation was primarily ascribed to the inhibition of Cu-ion transport and the higher oxidative stress [79]. Furthermore, Khan et al. [97] analyzed the MP vector effect using in vitro gut sacs and determined the intestinal destiny of combined effects of Ag and polyethylene MP beads. The result showed Ag was accumulated in intestinal compartments and neither MP nor Ag alone affects the trans-epithelial conveyance. However, the net result of the MP vector results in toxin forms in the intestine [97]. Conversely, Besseling et al. [99] studied the combined effects of PE and polychlorinated biphenyls (PCBs) transfer to Lugworm *Arenicola marina*. The quantified uptake fluxes result revealed PE has small effects on bioaccumulation and impacted feeding activity [98]. Furthermore, Qu et al. [99] inspected the degree of chiral product behaviors of the antidepressant venlafaxine and its metabolite in loach (Misgurnus anguillicaudatus) along with the effects of MPs toxicity, spread, and metabolism through a 40-day co-exposure to the loach. According to the study, MPs might be played a vector role to facilitate venlafaxine and O-desmethylvenlafaxine accumulated by loach. The bioaccumulation factor (BAF) for venlafaxine and O-desmethylvenlafaxine in loach tissue intensified more than 10 folds due to MPs presence. As a result, MPs assisted the transfer and bioaccumulation of pollutants to the liver causing delayed metabolism in an organism [99]. Similarly, Syberg et al. [100] evaluated the combined effects of PE microbeads (MP) and triclosan (TCS) in marine copepod *Acartia tonsa (Dana)*. The obtained data from the combined exposure of MP and hydrophobic organic (HOC) indicated that MP enhances the toxicity of TCS [100].

Zhu et al. [101] studied the combined toxicity effects of triclosan (TCS) with range polymers 74 mm PE, 74 mm PS, 74 mm PVC, and 1 mm PVC800 on marine phytoplankton. The associated triclosan (TCS) contaminant has shown growth inhibition, oxidative stress, superoxide dismutase (SOD), and malondialdehyde (MDA) microalgae *Skeletonema costatum*Skeletonema costatum. In addition single MPs also had noticeable inhibition effect in order of PVC800 > PVC > PS > PE. However, the combined toxicity effects of PVC and PVC800 in a mixture with TCS reduced over that of PE and PS due to the stronger adsorption ability of TCS on PVC and PVC800. Due to the minimum particle size of PVC800, the joint toxicity of PVC800 was still recorded as the highest toxic as compared to other polymers. Moreover, the higher reduction of SOD than MDA indicated that the physical impairment was more severely pronounced than intracellular damage [101]. Similarly, the toxicity of the pharmaceutical pollutants such as procainamide and doxycycline with assortments of MP with 1–5 μm diameter were examined on the microalga *Tetraselmis chuii*. For demonstration, a bioassay was exposed for 96 h to each pharmaceutical pollutants, MPs, and mixture of MP-pharmaceuticals. The toxicity of MPs and pharmaceuticals alone significantly reduced the growth rate and chlorophyll-a concentration. Moreover, toxicological mixture interaction of microplastics-pharmaceutical amplified the adverse toxic effect on *T. chuii* [102]. Comparatively, Batel et al. [103] explore the transference of MPs particles and concomitant POPs among different trophic levels through foodwebs with *Artemia sp. nauplii, and zebrafish (Danio rerio).* As of benzo[a]pyrene tracking fluorescence analyses, polymer particles in brine shrimp (Artemia sp.) nauplii were subsequently transferred with POPs to zebrafish via ingestion of exposed nauplii. Thus, food-borne MP-associated POPs can be desorbed in the intestine of fish which in turn transferred to the intestinal epithelium and liver. Yet, virgin particles without POPs load didn't cause visible physical detriment in the intestinal tracts of zebrafish. This implies MPs functioned as a vector facilitating the transfer of other contaminants to higher trophic levels in the food chain [103].

In addition, the joint toxicity of MP and ^{14}C-phenanthrene to *Daphnia magna* was examined for acute and long-term effects using 50 nm–10 μm nano plastic particles The common toxicity of 50-nm nano plastic particles and phenanthrene to *D. magna* exhibited an additive effect. In addition, 50-nm nanoparticles plastic indicated a significant dose-dependent effect. Furthermore, the incidence of nanoparticle plastic considerably boosted the bioaccumulation of phenanthrene-derived remains in the daphnid body and hindered the dissipation and alteration of phenanthrene in the medium. However, 10-μm MPs did not show major effects on the bioaccumulation, dissipation, and conversion of phenanthrene. The differences in toxicity effects between NPs sizes could be credited to greater adsorption of phenanthrene on 50-nm nano plastic particles than 10-μm MPs. In general, the findings pointed out that nano plastic particles are strong adsorbents for hydrophobic toxic pollutants and have a potential risk to aquatic ecology [104].

Additionally, plastic fragments are associated with numerous chemical contaminants identified to disturb the glands of aquatic organisms. The chronic exposure to no-plastic, virgin-plastic, and associated chemicals conducted using Japanese medaka (*Oryzias latipes*) in two monthly dietaries. The exposure to MPs (<1 mm) and supplementary chemicals raise endocrine disorder effects in fish. For example, a change in gene expression was detected in male fish exposed to the marine plastic, whereas the change in gene expression was detected in female fish exposed to the marine and virgin-plastic. For this reason, remarkable down-regulation of choriogenin (Chg H) gene expression was detected in males. In the same fashion remarkable down-regulation of vitellogenin (Vtg I), Chg H, and the estrogen receptor (ERα) gene expression was perceived in females. In addition, histological reflection has shown atypical rapid production of germ cells in one male fish from the marine plastic treatment. Overall, the assimilation of MPs fragments at environmentally relevant concentration may modify the endocrine system in adult fish [105]. Likewise, lugworms were exposed for 10 days to PVC sorbed with nonylphenol, triclosan, phenanthrene, and PBDE-47. The uptake of nonylphenol from PVC caused impaired inflammatory responses, and triclosan from PVC caused reduced survival and sediment burrowing activity of the worms [106]. Comparatively, lugworms were tested for toxicity effects of PCBs together with PS particles for 28 days. The result has shown joint toxic didn't have effects on survival, but PS caused reduced activity and weight loss, while PCBs were accumulated into tissues by a factor of 1.1–1.5 [107]. The combined toxicity effects of both the microplastic and associated are shown in Table 3.

7 Conclusion, Future Research, and Recommendation

Microplastics are a class of environmental pollutants and have become a prominent issue in the ecosystem. These trifling plastic fragments have been described to be extensively spread in near ranges of aquatic environments and biota everywhere the globe. Due to the ceaseless discharge of plastic waste, the plastic debris and the abundance of MP will continue in global waters. The incidence of MP in aquatic systems makes them vastly detected in aquatic organisms such as vertebrates, invertebrates, and primary producers. Present studies advocate that a large variety wide range of aquatic biota are prone to MP through ingestion and trophic transfer. The evidence from previous investigations revealed the presence of MPs after ingestion and tropical transfer. These microplastics are then accumulated in the gastrointestinal tracts and translocated to other organs such as the livers and gill of invertebrates. Because of their large surface area, microplastics also act as a potent vector role in transmitting hazardous chemicals, persistent organic contaminants, as well as leaches additive chemicals to aquatic biota. The exposure to MPs and associated contaminants of the aquatic biota and ecosystem affects the niche of a specific trophic level leading to changes in the ecosystem structure. Ecotoxicity figures show that MP exposure ultimately causes feeding loss, immature growth, reduced reproduction and offspring formation, etc., from producer to the consumer level. Its bioavailability to marine organisms seems rarely lethal, but chronic exposures to MPs in vertebrates can cause hepatic toxicity, reproductive toxicity, initiate immunological reactions, altered gene expressions, accumulation in gills, liver, and gut. Similarly, chronic exposure to

MPs in invertebrate triggered feeding rate reduction, accumulated in lipid storage droplets, increased enzyme activities, and alter feeding, energy assimilation, growth, and reproductive output. Besides, the chronic exposure to MPs in phytoplankton causes growth inhibition, reduces photosynthesis, and altered chloroplastic gene expression. Ultimately, microplastic exposure results in altered behavior, bioturbation, and carbon export. This causes motility, hiding responses, and predator-prey interactions resulting in an alteration of habitat structure which poses individual physiology with ecosystem function.

In this review paper, the occurrence, interaction, and toxicity behaviors of MP in the marine and freshwater aquatic biota were encapsulated, and the research advancements on the ecotoxicological effects of MPs and associated noxious waste on marine and freshwater aquatic organisms are discussed. Based on the researched indication presented in this paper, it is clear that there is a fundamental knowledge gap on the toxicity of MP infected biota on human health. Taking into consideration the imperative role of fish as a protein source for human beings, continual research investigations are extremely suggested to demonstrate the ecotoxicological effects of MPs on food webs from individual to population levels. This might also indorse the improvement and execution of proper strategies or protocols to cut the entry of MP into the aquatic environments that promote the protection of aquatic life.

Further research is required on the health impacts of MP on human beings due to ingestion through foodweb. In addition, the combined toxic effects of MPs and other associated environmental pollutants should also be studied extensively. Hence, it is a signal appeal for ecotoxicologists and ecologists to study the likely hostile effects of MPs at ecologically reported particle array and safeguard the ecological unit for sustainable development. Besides, detailed experimental researches on the ecotoxicological impacts through the food chain on human beings is required.

Tadele Assefa Aragaw and Bassazin Ayalew Mekonnen Conceptualization, Data gathering analysis, writing the original draft preparation, interpretation, Final reviewing, and editing.

Acknowledgements We would like to thank those organizations and scholars whom we have used previously findings related to this work. The authors also apologize to all intellectuals, and organizations whose involvement in the field of microplastic pollution and ecotoxicity may have been reviewed by the mistake or inadequately recognized.

Conflict of Interest The authors declare that they have no known competing financial interests or personal relationships that could have appeared to influence the work reported in this paper.

References

1. Geyer R, Jambeck JR, Law KL (2017) Production, use, and fate of all plastics ever made. Sci Adv 3:e1700782
2. Rochman CM, Browne MA, Halpern BS, Hentschel BT, Hoh E, Karapanagioti HK, Thompson CR (2013) Policy: classify plastic waste as hazardous. Nature 494:169–170. https://doi.org/10.1038/494169a

3. Pereao O, Opeolu B, Fatoki O (2020) Microplastics in aquatic environment: characterization, ecotoxicological effect, implications for ecosystems and developments in South Africa. Environ Sci Pollut Res 27:22271–22291. https://doi.org/10.1007/s11356-020-08688-2
4. Prokić MD, Radovanović TB, Gavrić JP, Faggio C (2019) Ecotoxicological effects of microplastics: examination of biomarkers, current state and future perspectives. TrAC Trends Anal Chem 111:37–46. https://doi.org/10.1016/j.trac.2018.12.001
5. Gregory MR (2009) Environmental implications of plastic debris in marine settings- entanglement, ingestion, smothering, hangers-on, hitch-hiking and alien invasions. Philos Trans R Soc B Biol Sci 364:2013–2025. https://doi.org/10.1098/rstb.2008.0265
6. Cole M, Lindeque P, Halsband C, Galloway TS (2011) Microplastics as contaminants in the marine environment: A review. Mar Pollut Bull 62:2588–2597. https://doi.org/10.1016/j.marpolbul.2011.09.025
7. Lusher A (2015) Marine anthropogenic litter: distribution, interactions and effects. In: Marine anthropogenic litter. Springer, Cham., pp 1–447. ISBN 9783319165103
8. Aragaw TA (2020) Surgical face masks as a potential source for microplastic pollution in the COVID-19 scenario. Mar Pollut Bull 159:111517. https://doi.org/10.1016/j.marpolbul.2020.111517
9. Sebille E, Van Gilbert A, Spathi C (2016) The ocean plastic pollution challenge: towards solutions in the UK. Grantham Inst Breifing Pap 19:1–16
10. Li WC, Tse HF, Fok L (2016) Plastic waste in the marine environment: a review of sources, occurrence and effects. Sci Total Environ 566–567:333–349. https://doi.org/10.1016/j.scitotenv.2016.05.084
11. Jambeck JR, Geyer R, Wilcox C, Siegler TR, Perryman M, Andrady A, Narayan R, Law KL (2015) Plastic waste inputs from land into the ocean. Science 347(80):768–771
12. Galloway TS, Cole M, Lewis C (2017) Interactions of microplastic debris throughout the marine ecosystem. Nat Ecol Evol 1:1–8. https://doi.org/10.1038/s41559-017-0116
13. Derraik JGB (2002) The pollution of the marine environment by plastic debris: a review. Mar Pollut Bull 44:842–852. https://doi.org/10.1016/S0025-326X(02)00220-5
14. Barletta M, Lima ARA, Costa MF (2019) Distribution, sources and consequences of nutrients, persistent organic pollutants, metals and microplastics in South American estuaries. Sci Total Environ 651:1199–1218. https://doi.org/10.1016/j.scitotenv.2018.09.276
15. Eriksen M, Lebreton LCM, Carson HS, Thiel M, Moore CJ, Borerro JC, Galgani F, Ryan PG, Reisser J (2014) Plastic pollution in the world's oceans: more than 5 trillion plastic pieces weighing over 250,000 tons afloat at sea. PLOS One 9:1–15. https://doi.org/10.1371/journal.pone.0111913
16. Thompson RC, Olson Y, Mitchell RP, Davis A, Rowland SJ, John AWG, McGonigle D, Russell AE (2004) Lost at sea: where is all the plastic? Science 304(80):838. https://doi.org/10.1126/science.1094559
17. Lebreton L, Slat B, Ferrari F, Sainte-Rose B, Aitken J, Marthouse R, Hajbane S, Cunsolo S, Schwarz A, Levivier A et al (2018) Evidence that the Great Pacific Garbage Patch is rapidly accumulating plastic. Sci Rep 8:1–15. https://doi.org/10.1038/s41598-018-22939-w
18. Horton AA, Walton A, Spurgeon DJ, Lahive E, Svendsen C (2017) Microplastics in freshwater and terrestrial environments: evaluating the current understanding to identify the knowledge gaps and future research priorities. Sci Total Environ 586:127–141. https://doi.org/10.1016/j.scitotenv.2017.01.190
19. Nelms SE, Galloway TS, Godley BJ, Jarvis DS, Lindeque PK (2018) Investigating microplastic trophic transfer in marine top predators. Environ Pollut 238:999–1007. https://doi.org/10.1016/j.envpol.2018.02.016
20. Welden NAC, Cowie PR (2016) Environment and gut morphology in fl uence microplastic retention in langoustine. Nephrops Norvegicus Environ Pollut 214:859–865. https://doi.org/10.1016/j.envpol.2016.03.067
21. Sathish MN, Jeyasanta I, Patterson J (2020) Occurrence of microplastics in epipelagic and mesopelagic fishes from Tuticorin, Southeast coast of India. Sci Total Environ 720:137614. https://doi.org/10.1016/j.scitotenv.2020.137614

22. Kumar VE, Ravikumar G, Jeyasanta KI (2018) Occurrence of microplastics in fishes from two landing sites in Tuticorin, South east coast of India. Mar Pollut Bull 135:889–894. https://doi.org/10.1016/j.marpolbul.2018.08.023
23. Su L, Deng H, Li B, Chen Q, Pettigrove V, Wu C, Shi H (2019) The occurrence of microplastic in specific organs in commercially caught fishes from coast and estuary area of east China. J Hazard Mater 365:716–724. https://doi.org/10.1016/j.jhazmat.2018.11.024
24. Ribeiro F, Garcia AR, Pereira BP, Fonseca M, Mestre NC, Fonseca TG, Ilharco LM, Bebianno MJ (2017) Microplastics effects in Scrobicularia plana. Mar Pollut Bull 122:379–391. https://doi.org/10.1016/j.marpolbul.2017.06.078
25. Steer M, Cole M, Thompson RC, Lindeque PK (2017) Microplastic ingestion in fish larvae in the western English Channel. Environ Pollut 226:250–259. https://doi.org/10.1016/j.envpol.2017.03.062
26. Desforges JPW, Galbraith M, Ross PS (2015) Ingestion of microplastics by zooplankton in the northeast pacific ocean. Arch Environ Contam Toxicol 69:320–330. https://doi.org/10.1007/s00244-015-0172-5
27. Van Cauwenberghe L, Claessens M, Vandegehuchte MB, Janssen CR (2015) Microplastics are taken up by mussels (Mytilus edulis) and lugworms (Arenicola marina) living in natural habitats. Environ Pollut 199:10–17. https://doi.org/10.1016/j.envpol.2015.01.008
28. Van Cauwenberghe L, Janssen CR (2014) Microplastics in bivalves cultured for human consumption. Environ Pollut 193:65–70. https://doi.org/10.1016/j.envpol.2014.06.010
29. Alomar C, Deudero S (2017) Evidence of microplastic ingestion in the shark Galeus melastomus Rafinesque, 1810 in the continental shelf off the western Mediterranean Sea. Environ Pollut 223:223–229. https://doi.org/10.1016/j.envpol.2017.01.015
30. Hoarau L, Ainley L, Jean C, Ciccione S (2014) Ingestion and defecation of marine debris by loggerhead sea turtles, Caretta caretta, from by-catches in the South-West Indian Ocean. Mar Pollut Bull 84:90–96. https://doi.org/10.1016/j.marpolbul.2014.05.031
31. Besseling E, Foekema EM, Van Franeker JA, Leopold MF, Kühn S, Bravo Rebolledo EL, Heße E, Mielke L, IJzer J, Kamminga P, et al (2015) Microplastic in a macro filter feeder: humpback whale Megaptera novaeangliae. Mar Pollut Bull 95:248–252. https://doi.org/10.1016/j.marpolbul.2015.04.007
32. Bravo Rebolledo EL, Van Franeker JA, Jansen OE, Brasseur SMJM (2013) Plastic ingestion by harbour seals (Phoca vitulina) in The Netherlands. Mar Pollut Bull 67:200–202. https://doi.org/10.1016/j.marpolbul.2012.11.035
33. Windsor FM, Tilley RM, Tyler CR, Ormerod SJ (2019) Microplastic ingestion by riverine macroinvertebrates. Sci Total Environ 646:68–74. https://doi.org/10.1016/j.scitotenv.2018.07.271
34. Su L, Nan B, Hassell KL, Craig NJ, Pettigrove V (2019) Microplastics biomonitoring in Australian urban wetlands using a common noxious fish (Gambusia holbrooki). Chemosphere 228:65–74. https://doi.org/10.1016/j.chemosphere.2019.04.114
35. Yuan W, Liu X, Wang W, Di M, Wang J (2019) Microplastic abundance, distribution and composition in water, sediments, and wild fish from Poyang Lake. China Ecotoxicol Environ Saf 170:180–187. https://doi.org/10.1016/j.ecoenv.2018.11.126
36. Xiong X, Zhang K, Chen X, Shi H, Luo Z, Wu C (2018) Sources and distribution of microplastics in China's largest inland lake–Qinghai Lake. Environ Pollut 235:899–906. https://doi.org/10.1016/j.envpol.2017.12.081
37. Bessa F, Barría P, Neto JM, Frias JPGL, Otero V, Sobral P, Marques JC (2018) Occurrence of microplastics in commercial fish from a natural estuarine environment. Mar Pollut Bull 128:575–584. https://doi.org/10.1016/j.marpolbul.2018.01.044
38. McNeish RE, Kim LH, Barrett HA, Mason SA, Kelly JJ, Hoellein TJ (2018) Microplastic in riverine fish is connected to species traits. Sci Rep 8:1–12. https://doi.org/10.1038/s41598-018-29980-9
39. Collard F, Gasperi J, Gilbert B, Eppe G, Azimi S, Rocher V, Tassin B (2018) Anthropogenic particles in the stomach contents and liver of the freshwater fish Squalius cephalus. Sci Total Environ 643:1257–1264. https://doi.org/10.1016/j.scitotenv.2018.06.313

40. Silva-Cavalcanti JS, Silva JDB, França EJ, de Araújo MCB, de Gusmão F (2017) Microplastics ingestion by a common tropical freshwater fishing resource. Environ Pollut 221:218–226. https://doi.org/10.1016/j.envpol.2016.11.068
41. Pazos RS, Maiztegui T, Colautti DC, Paracampo AH, Gómez N (2017) Microplastics in gut contents of coastal freshwater fish from Río de la Plata estuary. Mar Pollut Bull 122:85–90. https://doi.org/10.1016/j.marpolbul.2017.06.007
42. Hurley RR, Woodward JC, Rothwell JJ (2017) Ingestion of Microplastics by Freshwater Tubifex Worms. Environ Sci Technol 51:12844–12851. https://doi.org/10.1021/acs.est.7b03567
43. Su L, Cai H, Kolandhasamy P, Wu C, Rochman CM, Shi H (2018) Using the Asian clam as an indicator of microplastic pollution in freshwater ecosystems. Environ Pollut 234:347–355. https://doi.org/10.1016/j.envpol.2017.11.075
44. Zhang K, Xiong X, Hu H, Wu C, Bi Y, Wu Y, Zhou B, Lam PK, Liu J (2017) Occurrence and characteristics of microplastic pollution in Xiangxi Bay of Three Gorges Reservoir. China Environ Sci Technol 51:3794–3801. https://doi.org/10.1021/acs.est.7b00369
45. Faure F, Demars C, Wieser O, Kunz M, de Alencastro LF (2015) Plastic pollution in Swiss surface waters: nature and concentrations, interaction with pollutants. (Special Issue: Microplastics in the environment). Environ Chem 12:582–591
46. Andrady AL (2015) Persistence of plastic litter in the oceans. In: Bergmann M, Gutow L, Klages M (eds) Marine anthropogenic litter. Springer, Cham, pp 57–72. ISBN 9783319165097
47. Gewert B, Plassmann MM, Macleod M (2015) Pathways for degradation of plastic polymers floating in the marine environment. Environ Sci Process Impacts 17:1513–1521. https://doi.org/10.1039/c5em00207a
48. Browne MA, Dissanayake A, Galloway TS, Lowe DM, Thompson RC (2008) Ingested microscopic plastic translocates to the circulatory system of the mussel, Mytilus edulis (L.). Environ Sci Technol 43:5026–5031. https://doi.org/10.1021/es800249a
49. Moore CJ, Moore SL, Leecaster MK, Weisberg SB (2001) A comparison of plastic and plankton in the North Pacific Central Gyre. Mar Pollut Bull 42:1297–1300. https://doi.org/10.1016/S0025-326X(01)00114-X
50. Moore CJ (2008) Synthetic polymers in the marine environment: A rapidly increasing, long-term threat. Environ Res 108:131–139. https://doi.org/10.1016/j.envres.2008.07.025
51. Kühn S, van Franeker JA (2012) Plastic ingestion by the northern fulmar (Fulmarus glacialis) in Iceland. Mar Pollut Bull 64:1252–1254. https://doi.org/10.1016/j.marpolbul.2012.02.027
52. Eriksson C, Burton H (2003) Origins and biological accumulation of small plastic particles in fur seals from Macquarie Island. AMBIO A J Hum Environ 32:380–384
53. Farrell P, Nelson K (2013) Trophic level transfer of microplastic: Mytilus edulis (L.) to Carcinus maenas (L.). Environ Pollut 177:1–3. https://doi.org/10.1016/j.envpol.2013.01.046
54. Murray F, Cowie PR (2011) Plastic contamination in the decapod crustacean Nephrops norvegicus (Linnaeus, 1758). Mar Pollut Bull 62:1207–1217. https://doi.org/10.1016/j.marpolbul.2011.03.032
55. Setälä O, Fleming-Lehtinen V, Lehtiniemi M (2014) Ingestion and transfer of microplastics in the planktonic food web. Environ Pollut 185:77–83. https://doi.org/10.1016/j.envpol.2013.10.013
56. Bergmann M, Gutow L, Klages M (2015) Marine anthropogenic litter. ISBN 9783319165103
57. Ma H, Pu S, Liu S, Bai Y, Mandal S, Xing B (2020) Microplastics in aquatic environments: toxicity to trigger ecological consequences. Environ Pollut 261:114089. https://doi.org/10.1016/j.envpol.2020.114089
58. Wright SL, Thompson RC, Galloway TS (2013) The physical impacts of microplastics on marine organisms: a review. Environ Pollut 178:483–492. https://doi.org/10.1016/j.envpol.2013.02.031
59. Li J, Liu H, Chen JP (2018) Microplastics in freshwater systems: A review on occurrence, environmental effects, and methods for microplastics detection. Water Res 137:362–374. https://doi.org/10.1016/j.watres.2017.12.056

60. Lu Y, Zhang Y, Deng Y, Jiang W, Zhao Y, Geng J, Ding L, Ren H (2016) Uptake and accumulation of polystyrene microplastics in zebrafish (danio rerio) and toxic effects in liver. Environ Sci Technol 50:4054–4060. https://doi.org/10.1021/acs.est.6b00183
61. Qiao R, Sheng C, Lu Y, Zhang Y, Ren H, Lemos B (2019) Microplastics induce intestinal inflammation, oxidative stress, and disorders of metabolome and microbiome in zebrafish. Sci Total Environ 662:246–253. https://doi.org/10.1016/j.scitotenv.2019.01.245
62. Jin Y, Xia J, Pan Z, Yang J, Wang W, Fu Z (2018) Polystyrene microplastics induce microbiota dysbiosis and inflammation in the gut of adult zebrafish. Environ Pollut 235:322–329. https://doi.org/10.1016/j.envpol.2017.12.088
63. Watts AJR, Urbina MA, Goodhead R, Moger J, Lewis C, Galloway TS (2016) Effect of Microplastic on the Gills of the Shore Crab Carcinus maenas. Environ Sci Technol 50:5364–5369. https://doi.org/10.1021/acs.est.6b01187
64. Jeong CB, Won EJ, Kang HM, Lee MC, Hwang DS, Hwang UK, Zhou B, Souissi S, Lee SJ, Lee JS (2016) Microplastic Size-Dependent Toxicity, Oxidative Stress Induction, and p-JNK and p-p38 Activation in the Monogonont Rotifer (Brachionus koreanus). Environ Sci Technol 50:8849–8857. https://doi.org/10.1021/acs.est.6b01441
65. Jaikumar G, Brun NR, Vijver MG, Bosker T (2019) Reproductive toxicity of primary and secondary microplastics to three cladocerans during chronic exposure. Environ Pollut 249:638–646. https://doi.org/10.1016/j.envpol.2019.03.085
66. Gambardella C, Morgana S, Ferrando S, Bramini M, Piazza V, Costa E, Garaventa F, Faimali M (2017) Effects of polystyrene microbeads in marine planktonic crustaceans. Ecotoxicol Environ Saf 145:250–257. https://doi.org/10.1016/j.ecoenv.2017.07.036
67. Gambardella C, Morgana S, Bramini M, Rotini A, Manfra L, Migliore L, Piazza V, Garaventa F, Faimali M (2018) Ecotoxicological effects of polystyrene microbeads in a battery of marine organisms belonging to different trophic levels. Mar Environ Res 141:313–321. https://doi.org/10.1016/j.marenvres.2018.09.023
68. Romano N, Ashikin M, Teh JC, Syukri F, Karami A (2018) Effects of pristine polyvinyl chloride fragments on whole body histology and protease activity in silver barb Barbodes gonionotus fry. Environ Pollut 237:1106–1111. https://doi.org/10.1016/j.envpol.2017.11.040
69. Karami A, Romano N, Galloway T, Hamzah H (2016) Virgin microplastics cause toxicity and modulate the impacts of phenanthrene on biomarker responses in African catfish (Clarias gariepinus). Environ Res 151:58–70. https://doi.org/10.1016/j.envres.2016.07.024
70. Yin L, Chen B, Xia B, Shi X, Qu K (2018) Polystyrene microplastics alter the behavior, energy reserve and nutritional composition of marine jacopever (Sebastes schlegelii). J Hazard Mater 360:97–105. https://doi.org/10.1016/j.jhazmat.2018.07.110
71. Lei L, Wu S, Lu S, Liu M, Song Y, Fu Z, Shi H, Raley-Susman KM, He D (2018) Microplastic particles cause intestinal damage and other adverse effects in zebrafish Danio rerio and nematode Caenorhabditis elegans. Sci Total Environ 619–620:1–8. https://doi.org/10.1016/j.scitotenv.2017.11.103
72. Wong BBM, Candolin U (2015) Behavioral responses to changing environments. Behav Ecol 26:665–673. https://doi.org/10.1093/beheco/aru183
73. Rist SE, Assidqi K, Zamani NP, Appel D, Perschke M, Huhn M, Lenz M (2016) Suspended micro-sized PVC particles impair the performance and decrease survival in the Asian green mussel Perna viridis. Mar Pollut Bull 111:213–220. https://doi.org/10.1016/j.marpolbul.2016.07.006
74. Ziajahromi S, Kumar A, Neale PA, Leusch FDL (2017) Impact of microplastic beads and fibers on waterflea (ceriodaphnia dubia) survival, growth, and reproduction: implications of single and mixture exposures. Environ Sci Technol 51:13397–13406. https://doi.org/10.1021/acs.est.7b03574
75. Au SY, Bruce TF, Bridges WC, Klaine SJ (2015) Responses of Hyalella azteca to acute and chronic microplastic exposures. Environ Toxicol Chem 34:2564–2572. https://doi.org/10.1002/etc.3093
76. Murphy F, Quinn B (2018) The effects of microplastic on freshwater Hydra attenuata feeding, morphology & reproduction. Environ Pollut 234:487–494. https://doi.org/10.1016/j.envpol.2017.11.029

77. Bhattacharya P, Lin S, Turner JP, Ke PC (2010) Physical adsorption of charged plastic nanoparticles affects algal photosynthesis. J Phys Chem C 114:16556–16561. https://doi.org/10.1021/jp1054759
78. Besseling E, Wang B, Lürling M, Koelmans AA (2014) Nanoplastic affects growth of S. obliquus and reproduction of D. magna. Environ Sci Technol 48:12336–12343. https://doi.org/10.1021/es503001d
79. Qiao R, Lu K, Deng Y, Ren H, Zhang Y (2019) Combined effects of polystyrene microplastics and natural organic matter on the accumulation and toxicity of copper in zebrafish. Sci Total Environ 682:128–137. https://doi.org/10.1016/j.scitotenv.2019.05.163
80. Andrady AL (2011) Microplastics in the marine environment. Mar Pollut Bull 62:1596–1605. https://doi.org/10.1016/j.marpolbul.2011.05.030
81. Teuten EL, Rowland SJ, Galloway TS, Galloway TS (2007) Potential for plastics to transport hydrophobic contaminants potential for plastics to transport hydrophobic contaminants. ACS Publ 41:7759–7764
82. Teuten EL, Saquing JM, Knappe DRU, Barlaz MA, Jonsson S, Björn A, Rowland SJ, Thompson RC, Galloway TS, Yamashita R, et al (2009) Transport and release of chemicals from plastics to the environment and to wildlife. Philos Trans R Soc B Biol Sci 364:2027–2045. https://doi.org/10.1098/rstb.2008.0284
83. Ma P, Wei Wang M, Liu H, Feng Chen Y, Xia J (2019) Research on ecotoxicology of microplastics on freshwater aquatic organisms. Environ Pollut Bioavailab 31:131–137. https://doi.org/10.1080/26395940.2019.1580151
84. Antunes JC, Frias JGL, Micaelo AC, Sobral P (2013) Resin pellets from beaches of the Portuguese coast and adsorbed persistent organic pollutants. Estuar Coast Shelf Sci 130:62–69. https://doi.org/10.1016/j.ecss.2013.06.016
85. Imhof HK, Laforsch C, Wiesheu AC, Schmid J, Anger PM, Niessner R, Ivleva NP (2016) Pigments and plastic in limnetic ecosystems: A qualitative and quantitative study on microparticles of different size classes. Water Res 98:64–74. https://doi.org/10.1016/j.watres.2016.03.015
86. Napper IE, Bakir A, Rowland SJ, Thompson RC (2015) Characterisation, quantity and sorptive properties of microplastics extracted from cosmetics. Mar Pollut Bull 99:178–185. https://doi.org/10.1016/j.marpolbul.2015.07.029
87. Bakir A, Rowland SJ, Thompson RC (2014) Transport of persistent organic pollutants by microplastics in estuarine conditions. Estuar Coast Shelf Sci 140:14–21. https://doi.org/10.1016/j.ecss.2014.01.004
88. Betts K (2008) Why small plastic particles may pose a big problem in the oceans. Environ Sci Technol 42:8996. https://doi.org/10.1021/es802970v
89. Engler RE (2012) The complex interaction between marine debris and toxic chemicals in the ocean. Environ Sci Technol 46:12302–12315. https://doi.org/10.1021/es3027105
90. Ashton K, Holmes L, Turner A (2010) Association of metals with plastic production pellets in the marine environment. Mar Pollut Bull 60:2050–2055. https://doi.org/10.1016/j.marpolbul.2010.07.014
91. Ivleva NP, Wiesheu AC, Niessner R (2017) Microplastic in aquatic ecosystems. Angew Chemie Int Ed 56:1720–1739. https://doi.org/10.1002/anie.201606957
92. Fries E, Dekiff JH, Willmeyer J, Nuelle MT, Ebert M, Remy D (2013) Identification of polymer types and additives in marine microplastic particles using pyrolysis-GC/MS and scanning electron microscopy. Environ Sci Process Impacts 15:1949–1956. https://doi.org/10.1039/c3em00214d
93. Wagner M, Oehlmann J (2009) Endocrine disruptors in bottled mineral water: total estrogenic burden and migration from plastic bottles. Environ Sci Pollut Res 16:278–286. https://doi.org/10.1007/s11356-009-0107-7
94. Wagner M, Oehlmann J (2011) Endocrine disruptors in bottled mineral water: estrogenic activity in the E-Screen. J Steroid Biochem Mol Biol 127:128–135. https://doi.org/10.1016/j.jsbmb.2010.10.007

95. Kim D, Chae Y, An YJ (2017) Mixture toxicity of nickel and microplastics with different functional groups on daphnia magna. Environ Sci Technol 51:12852–12858. https://doi.org/10.1021/acs.est.7b03732
96. Barboza LGA, Vieira LR, Branco V, Figueiredo N, Carvalho F, Carvalho C, Guilhermino L (2018) Microplastics cause neurotoxicity, oxidative damage and energy-related changes and interact with the bioaccumulation of mercury in the European seabass, Dicentrarchus labrax (Linnaeus, 1758). Aquat Toxicol 195:49–57. https://doi.org/10.1016/j.aquatox.2017.12.008
97. Khan FR, Boyle D, Chang E, Bury NR (2017) Do polyethylene microplastic beads alter the intestinal uptake of Ag in rainbow trout (Oncorhynchus mykiss)? Analysis of the MP vector effect using in vitro gut sacs. Environ Pollut 231:200–206. https://doi.org/10.1016/j.envpol.2017.08.019
98. Besseling E, Foekema EM, Heuvel-Greve V, Den JM, Koelmans AA (2017) The Effect of Microplastic on the Uptake of Chemicals by the Lugworm Arenicola marina (L.) under Environmentally Relevant Exposure Conditions. Environ Sci Technol 51:8795–8804. https://doi.org/10.1021/acs.est.7b02286
99. Qu H, Ma R, Wang B, Yang J, Duan L, Yu G (2018) Enantiospecific toxicity, distribution and bioaccumulation of chiral antidepressant venlafaxine and its metabolite in loach (Misgurnus anguillicaudatus) co-exposed to microplastic and the drugs. J Hazard Mater 370:203–211. https://doi.org/10.1016/j.jhazmat.2018.04.041
100. Syberg K, Nielsen A, Khan FR, Banta GT, Palmqvist A, Jepsen PM (2017) Microplastic potentiates triclosan toxicity to the marine copepod Acartia tonsa (Dana). J Toxicol Environ Heal Part A Curr 80:1369–1371. https://doi.org/10.1080/15287394.2017.1385046
101. Zhu Z-L, Wang S-C, Zhao F-F, Wang S-G, Liu F-F, Liu G-Z (2019) Joint toxicity of microplastics with triclosan to marine microalgae Skeletonema costatum. Environ Pollut 246:509–517. https://doi.org/10.1016/j.envpol.2018.12.044
102. Prata JC, Lavorante BRBO, Maria da M, da C, Guilhermino L (2018) Influence of microplastics on the toxicity of the pharmaceuticals procainamide and doxycycline on the marine microalgae Tetraselmis chuii. Aquat Toxicol 197:143–152. https://doi.org/10.1016/j.aquatox.2018.02.015
103. Batel A, Linti F, Scherer M, Erdinger L, Braunbeck T (2016) Transfer of benzo[a]pyrene from microplastics to Artemia nauplii and further to zebrafish via a trophic food web experiment: CYP1A induction and visual tracking of persistent organic pollutants. Environ Toxicol Chem 35:1656–1666. https://doi.org/10.1002/etc.3361
104. Ma Y, Huang A, Cao S, Sun F, Wang L, Guo H, Ji R (2016) Effects of nanoplastics and microplastics on toxicity, bioaccumulation, and environmental fate of phenanthrene in fresh water. Environ Pollut 219:166–173. https://doi.org/10.1016/j.envpol.2016.10.061
105. Rochman CM, Kurobe T, Flores I, Teh SJ (2014) Early warning signs of endocrine disruption in adult fish from the ingestion of polyethylene with and without sorbed chemical pollutants from the marine environment. Sci Total Environ 493:656–661. https://doi.org/10.1016/j.scitotenv.2014.06.051
106. Browne MA, Niven SJ, Galloway TS, Rowland SJ, Thompson RC (2013) Microplastic moves pollutants and additives to worms, reducing functions linked to health and biodiversity. Curr Biol 23:2388–2392. https://doi.org/10.1016/j.cub.2013.10.012
107. Besselincg E, Wegner A, Foekema EM, Van Den Heuvel-Greve MJ, Koelmans AA (2013) Effects of microplastic on fitness and PCB bioaccumulation by the lugworm Arenicola marina (L). Environ Sci Technol 47:593–600. https://doi.org/10.1021/es302763x
108. Avio CG, Gorbi S, Milan M, Benedetti M, Fattorini D, D'Errico G, Pauletto M, Bargelloni L, Regoli F (2015) Pollutants bioavailability and toxicological risk from microplastics to marine mussels. Environ Pollut 198:211–222. https://doi.org/10.1016/j.envpol.2014.12.021

Microplastic Pollution in Marine Environment: Occurrence, Fate, and Effects (With a Specific Focus on Biogeochemical Carbon and Nitrogen Cycles)

Bozhi Yan, Qing Liu, Jingjing Li, Chunsheng Wang, Yanhong Li, and Chunfang Zhang

Abstract The pollution of microplastics is becoming increasingly serious, and rising evidence shows that the marine environment, especially sediments are major sinks of these plastics. So far, microplastic particles have been reported as widespread in large quantities in various water body and sedimentary environments such as beach, shallow coastal area, estuary, fjord, continental shelf environments, and deep-sea environments. Moreover, recent studies showed that the existence of microplastics would influence the structure and function of marine environmental microbial communities, thereby affecting the nitrogen/carbon cycling processes in marine environment especially sediments. Considering increasing microplastic pollution in marine environment especially sediment, the impact of plastics on marine ecosystems and biogeochemical cycling deserves in-depth investigation. Therefore, in the present study, information on occurrences and fate of microplastic particles in different marine environments, and their effects, mainly on marine ecosystems and biogeochemical carbon and nitrogen cycles, is reviewed.

Keywords Microplastics · Sediment · Transportation · Biogeochemical cycle · Plankton · Benthos · Microbial communities

---

B. Yan · Q. Liu · Y. Li · C. Zhang (✉)
College of Environmental Science and Engineering, Guilin University of Technology, Guilin 541006, China
e-mail: 0014105@zju.edu.cn; zhangcf@zju.edu.cn

J. Li · C. Zhang
Ocean College, Zhejiang University, Zhoushan 316021, Zhejiang, China

C. Wang
Second Institute of Oceanography, State Oceanic Administration, Hangzhou, China

S. S. Muthu (ed.), *Microplastic Pollution*, Sustainable Textiles: Production, Processing, Manufacturing & Chemistry, https://doi.org/10.1007/978-981-16-0297-9_4

1 Introduction

Nowadays, human life is full of plastics, and their production and consumption have rapidly increased since the 1950s, Millions of tons of plastic products are being produced worldwide every year [1]. Since the beginning of the last century, the output of plastics has increased year by year with the development of science and technology, and the annual production reaches almost 360 million tons in 2018 [2]. It is being widely used in industry, agriculture, aerospace, daily life, and other fields because of its low cost, strong plasticity, high strength, and ease of manufacture. In the past few decades, plastic pollution in the ecological environment has dramatically increased due to its wide application in many different fields and its properties of persistence and refractory. It is estimated that at least 4.8–12.7 million t of plastic waste enter the marine environment from the land each year through surface runoff and other means [3], causing serious threats to the marine ecological environment. Plastics in the ocean can be gradually decomposed through ultraviolet radiation, biodegradation, and oxidation and eventually disintegrate into small debris. This study reviewed the occurrences and fate of microplastics in different marine environments, and their effects, mainly on sedimentary microbial ecosystems and biogeochemical carbon and nitrogen cycles, as shown in the roadmap (Fig. 1).

1.1 Concept and Composition of Microplastics

Microplastic pollution is becoming more and more widespread on a global scale and causing irreversible damage to the geochemical cycle of the ecological environment. In recent years, the effect of microplastics on the marine ecological system has been widely investigated by large numbers of researchers, such as the distribution and abundance of microplastics in the marine environment, and the ecotoxicological effects of microplastics on marine organisms.

The concept of "microplastics" was initially proposed by Thompson et al. [4] to describe plastic particles. In some cases, plastic based on its size is also categorized as microplastics (<2 mm), mesoplastics (2 mm–2 cm), and macroplastics (>2 cm) [5]. At present, there is still no standardized criteria for the size of microplastics, but it usually refers to microplastics with a radius less than 5 mm. Microplastics are also referred to as "$PM_{2.5}$ in the ocean" because of their tiny size that is hard to observe with bare eyes. According to Anderson et al. [6], microplastic types are generally classified into five categories: fragments, micro-pellets, fibers, films, and foam. Whereas plastic based on its basic type is separated as polyethylene terephthalate (PET), polyethylene (PE), polyvinylchloride (PVC), polypropylene (PP), polystyrene (PS), and others [7].

Microplastics pose serious threats to the marine ecological environment on a global scale. Nowadays, extensive studies investigated the distribution and abundance of plastic debris in different marine environments [8–10]. As abundance is commonly used to evaluate the distribution of microplastics in the marine environment such as

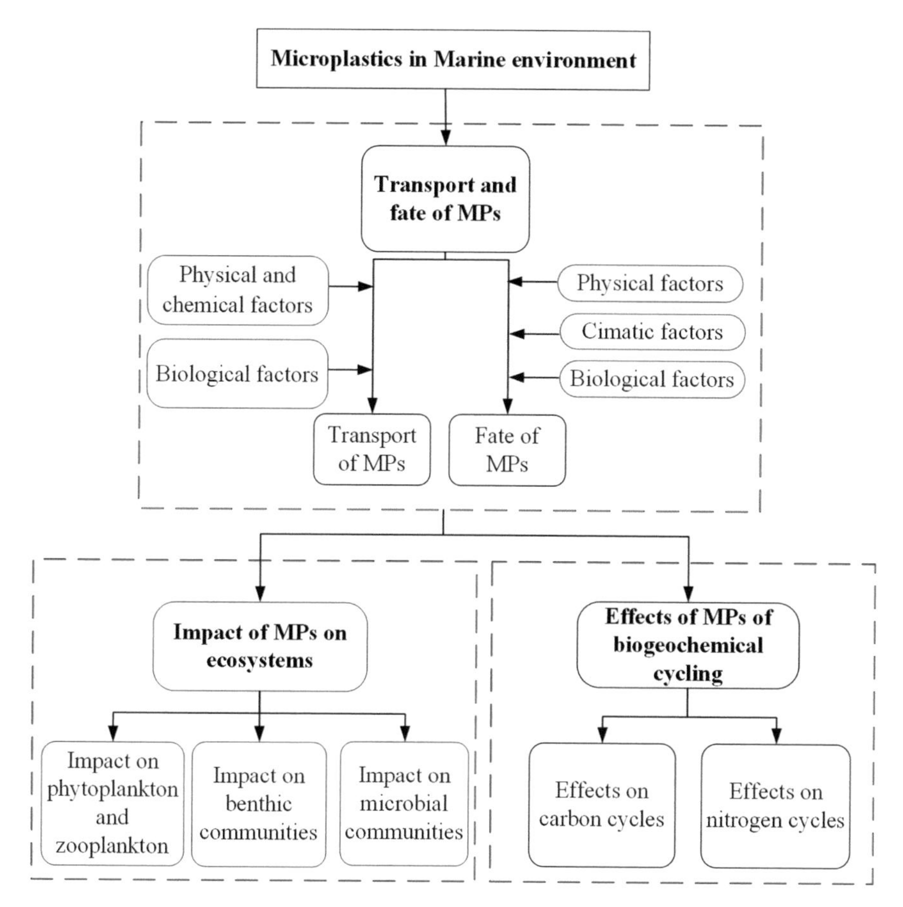

Fig. 1 Roadmap of chapter

surface water and sediments, the quantity of particle is generally taken as the basic unit, i.e., the distribution of microplastics in surface water is usually described as items m^{-3}, items L^{-1}, and items Km^{-2}, while items kg^{-1}, items m^{-2} are usually used to describe the microplastics detected in the sediment.

1.2 Sources of Microplastics

Microplastics can be divided into two categories according to their source, namely primary microplastics and secondary microplastics [6]. The former refers to the plastic particles that are intentionally created for some specific purposes [11], which mainly include resin particles and microbeads [12]; the latter refers to small pieces of plastic formed through the processing of physical, chemical, biological, and non-biological decomposition, etc., from the large pieces of plastics entering the environment [13]. For example, many industrial raw materials, household articles, plastic garbage, packaging materials, and other products produced by human activities contain plastic components. All these could serve as possible sources of microplastic through fragmentation or degradation.

Microplastics in the marine environment mainly comes from surface runoff from terrestrial rivers. The plastic garbage discarded along the coast enters into the marine environment through the runoff and ocean currents. Garbage stacking and wastewater treatment near the shoreline would contribute to the marine microplastic accumulation. In addition, near-marine aquaculture plastic products are also an important source of marine microplastics.

The public has come to realize that the large amount of microplastics in the ocean is a global concern. As the main pollution area, microplastics could be found in various marine environments, from coastal to deep-sea environments. At the very beginning, most researches focused on microplastic distribution in surface water. Later, researchers found that in spite of the property of buoyancy, microplastics could be still transported into sediments after being ingested by a marine organism or colonized by microorganisms [10, 14].

More and more studies have shown that the level of microplastic in different sediment environments (such as coastal, nearshore, offshore, deep sea, etc.) is relatively high. It is now widely accepted that marine sediments are considered to be long-term microplastic deposits [15], therefore, the effects of microplastics on sedimentary communities and geochemical cycles are worth studying.

2 Environmental Fate and Transport of Microplastics

2.1 The Transport of Microplastics in the Marine Environment

In order to understand the fate of microplastics, it is important to know their behaviors of transport. It is well known that air, water, and living things are the transmission factors of microplastics, among which the water environment transports the most microplastics. In aquatic environments, abiotic factors such as water flow, currents, depth, wind, surface area, and density of suspended particles would all affect the transport of microplastics. Microplastics generally float on the overlying water due to their lower density and higher buoyancy and exist in the ocean surface. As a carrier of microplastics, surface water can concentrate microplastics more easily than the bottom water [16]. Generally, the smaller the size of microplastics, the greater the buoyant force of microplastics, and more easily rising to the surface of the sea. Land-based microplastics enter the ocean through rivers and it is often called advection transport [16].

In addition to the above factors, the different terrain environment will also affect the transport of microplastics. Many studies have shown that sea winds in open coastal zones cause strong dispersal and transport of microplastics to the deep sea, while in semi-enclosed areas they cause strong trapping and enrichment in sediments [17].

Compared to the overlying water, there are large amounts of microplastics in sediments at the bottom of the ocean. Many studies have shown that microplastics in marine sediments account for 80% of the total content of marine microplastics. Numerous studies have generally agreed that sediment is the ultimate destination of microplastics [15]. Compared with the commonly believed transport of microplastics in an aquatic environment, the increasing number of studies have shown that atmospheric transport [18] has become one of the important means of migration and transport of microplastics to remote marine environments, such as deep sea or polar regions, and will have a profound impact on the marine ecological environment [19].

The transport of microplastics in the ocean depends not only on natural factors but also on biological factors. In earlier studies, seabirds have been identified as vectors of chemical pollutants from the terrestrial environment to the marine environment. More recently, Radisic et al. [17] reported that seabirds may be vectors of microplastic contamination in marine sediment environments by feeding on microplastic-enriched organisms and then sinking microplastic-containing feces into the sediment by excreting them [17].

Microplastics not only have biotoxicity, bio-accumulation, but also have unique properties. They are carriers of persistent organic pollutants or biological pathogens [20], thus causing harm to the marine environment. Recently, several studies have shown that microplastics may be the carrier of harmful pollutants (i.e., hydrophobic organic compounds, chemical additives, pathogens) [21]. When ingested by low-level microorganisms, microplastics can be enriched into a higher level of marine organisms through the food chains and eventually enter the human body, causing harm to human health. In addition, the additives contained in microplastics will also be released into the marine environment during the aging process, causing physiological toxicity to marine organisms [22]. The additive, which is in microplastics, interacts with the organic pollutants that are adsorbed on its surface, enhancing toxicity. Some studies have shown that microplastics can transfer pollutants into organisms [23]. Bowley et al. [20] have reported that marine microplastics can not only carry potentially toxic fish pathogens, but also carry potential human pathogens and act as carriers for these pathogens to be transported in the marine environment [20]. Studies have presented that the toxic pollutants polychlorinated biphenyl show relatively obvious adsorption affinity on PET and PS, and are easy to enter the marine ecosystem through microplastics.

2.2 *The Fate of Microplastics in the Marine Environment*

There is no consensus on where microplastics will end up in the ocean. Generally, we believed that microplastics in the marine environment will go through a series of migration and transformation, and the high-density microplastics will gradually be deposited in coastal areas under the drive of waves and tides. Microplastics with low density will float or be suspended in the seawater and will diffuse under the driving forces of ocean currents, tides, wind waves, tsunamis, and other dynamic processes. Under the long-term action of the marine environment, the surface characteristics of hydrophobic microplastics become complicated, and it is easy to absorb some organic pollutants and metal chemical pollutants, and some clay particles, organic debris, seaweed, microorganisms, etc., will also be attached. These processes will increase the density of microplastic particles or change their surface characteristics, prompting them to settle. The latest reports also confirmed that microplastic pollution is widespread in the sediment environment [24].

Data show that 70% of marine microplastics are deposited in seafloor sediments, 15% of microplastics float in coastal areas, and the remaining 15% float in surface waters. But over time, the microplastics that enter the ocean eventually settle into sediments on the seafloor. Recent estimates suggested that 4.85 trillion microplastic particles are floating in the world's oceans [25]. Francesca and Ryan [26] have shown that microplastics with a low density float on the surface of seawater, but due to their small size, they are easily swallowed by plankton and enter the seafloor with their feces [26]. Smaller microplastics tend to sink faster than larger microplastics. Some researches suggested that if microplastics are not transported into the ocean for a long time, all the microplastics in the oceans will sink into the sediments at the bottom of the sea. Nearly 8 million t of plastic waste is discharged into the ocean every year. The accumulation rate of microplastics in the ocean is much higher than the removal rate. At present, many early microplastics accumulated in the ocean are gradually sinking and will eventually accumulate in the deep sea and seabed sediments, which will pose serious ecological risks to the benthic ecosystems. Tables 1 and 2 summarized the distribution of microplastics in marine water, sediments, and marine organisms all over the world in recent years in terms of morphology and abundance.

3 The Impact of Microplastics on Ecosystems

Microplastics are widely existing in the marine ecosystem. Although extensive studies have been carried out to investigate the abundance and distribution of microplastics in the marine environment, few studies involve the influence of microplastics on the marine ecosystem function and the geochemical cycle. Therefore, it is necessary to study the effects of microplastics in this aspect.

Table 1 Form and abundance of microplastics in various marine waters and marine sediments

	Locations	Dominant types of microplastics	Microplastics size (mm)	Abundance/Concentration	References
MPs in water samples of global regions	Geoje and Jinhae Bays	Fiber, Hard plastic, Paint particles, Styrofoam	<1.5	1.92 ± 1.84 p/m^3 (before rainy season); 5.51 ± 11.24 p/m^3 (after rainy season)	[27]
	Yangtze Estuary	Fibers, granules, films	0.5–5	4137.3 ± 2461.5 p/m^3 (Estuarine); 0.167 ± 0.138 p/m^3 (Sea)	[28]
	Tokyo Bay, Suruga Bay, Ise Bay, and Seto Inland Sea	Undegraded spherical microbeads; Spherical PE particles	0.3–5	0.161–0.446 p/m^3	[29]
	Richard's Bay Harbour, Durban Harbour	Fibers	0.063–5	413.3 ± 77.53 p/m^3 (Richard's); 1200 ± 133.2 p/m^3 (Durban)	[30]
	South-eastern coastline	Fibers	0.080–5	257.9 ± 53.36–1215 ± 276.7 p/m^3	[31]
	San Francisco Bay	Fragments, Fibers	0.355–0.999	0.15–2 p/m^2	[32]
	Louisiana Coast	Fibers	<0.335	5.0–18.4 p/m^3	[33]
	South Adriatic Sea	PE	0.2–0.5	0.04–4.65 p/m^2	[34]
MPs in sediment samples of global regions	Guanabara Bay	PP, nylon, PVA	<5	12–1300 p/m^2	[35]
	Bays in Berlayar Creek	Fibers, films, granules	<0.040	36.8 ± 23.6 p/kg	[36]
	Tromsø	Polyester, PP, PE	<1	72 ± 24 p/kg	[37]
	Balearic Islands	Fragments (UDP)	0.5–2	900 ± 100 p/kg	[13]
	Ma'an Archipelago	PE, PP	<1	30.0 ± 0.0–80.0 ± 14.1 p/m^2	[10]

(continued)

Table 1 (continued)

	Locations	Dominant types of microplastics	Microplastics size (mm)	Abundance/Concentration	References
	Belgian Continental Shelf, the North Sea	Spheres (UDP), fibers	0.355–0.499	0–97.2 p/kg	[38]
	Coast in Southern Yellow Sea and East China Sea	PE,PET	<1	134 ± 6 p/kg	[39]
	Western Pacific Ocean	PP-PE,PET	<1	0–1042 p/kg	[9]
	Shalun Beach, Baishawan Beach, Waimushan Beach and Fulong Beach	Fragment (UDP), foam, pellets, fibers	–	40–5320 p/m^3	[40]
	Bays in China	PP, PE	–	97.5 ± 157.4 p/m^2	[29]
	Baynes Sound and Lambert Channel, British Columbia	Fragment (UDP), fibers, fragments	–	up to 25,000	[41]
	Lagoon-Channel of Bizerte	Fragment (UDP), fibers, fragments	0.3–5	3000–18,000	[42]
	Canadian Lake Ontario nearshore, tributary and beach sediments	Fragment (UDP), fibers, fragments	<2	20–27,830	[43]
	The tidelands of Osaka Bay in Japan	PVC, PS, PVA, PE	0.3–5	79–890 P/kg	[44]
	Coastal sediments of Jakarta Bay	Fragment (UDP), fibers, fragments, pellets	1–5	18,405–38,790 p/kg	[45]

Note Standardized with data to maintain consistency across all data. The symbol "–" indicates that the data is not available. Units p/m^2 or p/m^3 represent the amount of microplastics per m^2 or m^3; p/kg or p/beach units represent the amount of microplastics per kg of sample or beach, respectively. PA: Polyamide; UDP (Undetermined plastic particles): Undetermined plastic particles; PEVA: Polyethylene acetate; LDPE: low-density polyethylene; PVA: Polyethylene alcohol; PP: polypropylene; PS: Polystyrene; PVC: Polyvinyl chloride

Table 2 Distribution of microplastics in marine organisms

Biological sample	Dominant Type/Composition of microplastics	Microplastics size (mm)	Abundance (p/N)	Common sample	Sampling area	References
Mussel	Fiber, fragment/PE, PP, PS	0.005–5	0.2 ± 0.3– 9.2 p/g	Soft tissue, intestine	China, Brazil, UK, French, Italy, etc.	[46–51]
			A. marina: 1.2 ± 2.8 p/g			
Zooplankton	Fibrous/PE	0.10–0.14	0.6–4.1 p/m^3 (505 μm sampling net)	Individual	Northern South China Sea; Portuguese	[52, 53]
Fish	Fiber, fragment, filament, sphere, microbead, pellet/PE, PP, rayon	1 nm–5 mm	0.146–21.8	Tissue, Digestive tract, Individual	Australia, Spanish, Scotland, Turkey, USA, China	[54–63]
Seabird	Fragment, filament, thread	0.4–42.4 (Length); 0.01–19.5 (Width)	9.99–11.6	Gular pouch	Southern Labrador Sea (Canada); East Greenland	[64, 65]
Whale (Fulmars, Shearwaters)	Fragment	–	Fulmars: 19.5 Shearwaters: 13.3		Pacific Ocean beaches of Tillamook and Clatsop counties	[66]
Turtle	PEAA, PVA	0.45–2.95	–		Great Barrier Reef	[67]

3.1 Impacts on Marine Phytoplankton and Zooplankton

Phytoplankton are the main producers in the ocean, whose primary productivity accounts for about 80% of the planet's total oxygen production. The change of phytoplankton is closely related to the marine geochemical cycle, among which microalgae are the most widespread ones. Microalgae belong to autotrophic phytoplankton which is ubiquitous in the marine ecosystem, rich in nutrients, and highly photosynthetic. Microalgae are the basic components of aquatic food webs and the most essential plants in aquatic ecosystems [68]. A number of studies had confirmed the fact that microplastics do interact with microalgae, and this affects their fates.

According to the existing studies, the influence of microalgae on microplastics is mainly divided into two aspects, one is the change in the properties of plastic polymers/or biodegradation; the other is the change in polymer density and sinking behavior (Nava and Leoni). Some studies had also suggested that the biological adhesion process of microplastics may affect its properties, with its adsorption capacity seems to be accelerated [69, 70]. One study showed that phytoplankton attached to PE microplastics led to changes in the physicochemical properties of the microplastics, and ultimately gave rise to changes in the adsorption performance of the microplastics on copper and Tetracycline. When the growth of phytoplankton is restricted, a variety of algae can secrete viscous substances to polymerize with microplastics and form algal clusters [71], which can not only change the density and distribution of microplastics in the ocean, but also promote the low-density plastic particles to sink into the sediment.

Similarly, microplastics can pose harmful effects on phytoplankton. Microplastics may have toxic effects on marine phytoplankton due to their hydrophobic property which can serve as carriers of organic pollutants and heavy metals [72]. Microalgae are essential primary producers of the aquatic ecosystems [73] and the toxicity of microplastics exposed to them may affect the entire marine food web.

There are many kinds of microplastics entering the ocean. Low microplastic concentration has little influence on the growth of phytoplankton, especially algae. So its impacts are almost negligible [74]. However, a high concentration of microplastics would pose serious adverse effects on the growth and development of phytoplankton. It has been reported that under the circumstance of high polystyrene concentration, the production of reactive oxygen species (ROS) in chlorella cells would be accelerated, thus increasing the degree of cell apoptosis and adversely affecting the production of chlorella [75].

In addition to various polymer types and additives within different doses of microplastics, other characteristics such as the size of microplastics may also be critical to marine phytoplankton. A growing number of studies has shown a link between the particle size of microplastics and toxicity, and it is generally considered that the toxicity of microplastics to microalgae increases as the size of microplastics decreases [76]. Tiny particles may inhibit the growth of microalgae more easily by attaching to the surface of algal cells. For example, inducing shading, blocking algal pores or gas exchange, and embedding in microalgal cells [77]. The ingestion, metabolism, development, and reproduction of phytoplankton are all influenced by microplastic contamination. The toxicity increases significantly with the decrease of microplastics size.

The relationship between microplastics and microalgae interacted, some studies also reported that microplastics can also affect the photosynthesis of algae. The chlorophyll content and photosynthetic efficiency of algae are reduced under the exposure of microplastics [78].

Mao et al. have reported that exposure to polystyrene (PS) microplastics causes an oxidative stress response, cell membrane damage, and decreased photosynthetic activity of chlamydomonas nucleolus [79]. Ting et al. have also shown that micro polyvinyl chloride has adverse effects on chlorophyll content and photosynthetic efficiency of phytoplankton (dinoflagellates) [80]. Besides, when exposed to microplastics, the photosynthetic efficiency of phytoplankton such as Dunaliella declined by 45% [80]. Bhattacharya et al. reported that the content of chlorophyll in algal cells would reduce when low-concentration polystyrene microplastics were exposed to chlorella and phyllostrella, thereby reducing the efficiency of photosynthesis [75]. It increases the production of ROS in algal cells and causes apoptosis. Other studies reported that the chlorophyll synthesis content and photosynthetic efficiency are significantly decreased when polystyrene is exposed to phytophthora [81].

As the primary consumer of phytoplankton, zooplankton is also threatened by microplastic pollution. Microplastics might be mistakenly taken and ingested by any zooplankton, including rotifers, copepods, and polychaete worms [82]. Once ingested, microplastics can be transferred through the food chain from small zooplankton to larger zooplankton [83].

Besides, microplastics attached to phytoplankton could be eaten by zooplankton, which would be eventually transported by fecal particles to deep seabed sediments. However, some researches have shown that when fecal particles contain microplastics, the subsidence rate of fecal particles decreases by 1.35 times, and they are more likely to be broken [84], which makes microplastics more likely to float in the upper water and has an impact on the degradation process of microplastics.

3.2 *Impacts on Marine Benthic Communities*

Benthic organisms play a vital role in the sediment environment, and different benthic organisms can maintain the geochemical cycling process in sediments such as the nitrogen cycle, by modifying sediments and regulating microbial activity [85]. Benthic organisms with different functional characteristics can participate in the maintenance of ecosystem functions, thereby improving the resilience of the ecosystem in the sediment environment [86]. Therefore, the effect of microplastics on benthic organisms is of great significance to the marine ecological environment.

The toxicity of microplastics to benthic organisms depends on whether they can be enriched in benthic organisms. According to documented studies, the excretion of benthic organisms such as mussels will increase with the concentration of microplastics in the sediments [87]. Besides, Sinja et al. reported that microplastic particles can be deposited in mussels and cannot be effectively digested and broken down [88].

Microplastics also alter the feeding behavior of benthic organisms. The inhomogeneous particle size of microplastics may affect the feeding of marine filter-feeding bivalves and may further lead to different physiological effects [39]. Studies had shown that microplastics cannot be ingested by bivalve mollusks if they are separate in the environment. Bivalves depend on the type and size of microplastics to determine whether they are suitable for food. Since benthic organisms can selectively ingest microplastics, the physical and chemical properties of microplastic particles could influence their existence in the intestines of bivalves, such as the type and size of the microplastics [89]. If microplastics are mistakenly eaten by mussels with nutritious foods, they tend to accumulate in their bodies, and if microplastics are eaten by mussels alone, on the contrary, they are excreted, and the results show that mussels can distinguish between nutritious foods and uneatable suspended plastic particles [76].

Microplastics have different effects on numerous benthic organisms which result in the destruction of their physiological functions. Benthic invertebrates, especially those species that absorb benthic sediments such as angelfish (blackworms), are usually found in contaminated sediments and their bodies contain large quantities of microplastics [90]. As a common benthic organism, midges are also affected by the accumulation of microplastics. The accumulation of microplastics may not only inhibit the metabolism, nutrition, immunity, and development time of midges, but also lead to the changes of intestinal microflora of midges.

3.3 Effects of Microplastics on Microbial Diversity, Community Structure, and Function

The existence of microplastics has a significant impact on microorganisms, it could change the microbial diversity and community structure, and ultimately affecting the function of microbial communities. Li et al. drew the conclusion that the existence of microplastics will affect the structure and composition of the microbial community as well as the function of the microbial community, potentially affecting the ecological function of the microbial community in the aquatic ecosystem by applying multiple analysis [91].

Microplastics have a large specific surface area, which makes them an ideal habitat for microorganisms to form biofilms. As a unique habitat, the adhesion and formation of biofilm on the microplastic would be affected by the plastic surface, the surrounding environment, seasonal variation, as well as geographical location [92]. Harrison et al. reported that microorganisms have the ability to rapidly attach to LDPE microplastics in coastal marine sediments [93]. In addition, Miao et al. suggested that microplastics, as an ideal plastisphere, may be more abundant in microbial community composition than other natural substrates [94]. Moreover, the microbial colonization of plastics can also be considered as biofouling. The formation of biofouling may lead to changes in the buoyancy density of the polymer and promote the migration of microplastics from the ocean surface to deeper water columns and sediments [95].

Microplastics are sometimes used not only as a carbon source, but also absorbing organic particles from the ambient environment to provide a carbon source for microorganisms with carbon metabolism functions [8]. According to relevant reports, many bacterial strains have the ability to degrade microplastics. It was reported that during their degradation process, microplastics can produce toxic substances, such as phthalates, which can be toxic to microorganisms [78]. Generally, in the process of biodegradation of microplastics, the first and foremost important step is the attachment and colonization of the microorganisms on the microplastics, followed by the formation and utilization of microplastics as a carbon source for growth (Nava and Leoni).

In the sedimentary environment, the influences of different microplastics on microbial diversity are greatly diverse. The study shows that the microbial diversity is rich under the treatment of polylactic acid microplastics in the coastal salt marsh sediment environment, while the microbial community diversity is significantly reduced in the presence of PE microplastics under the same environment [96]. Thus, further study is needed to explore the influence of microplastics on specific functions of microorganisms, as well as the mechanisms involved.

4 Effects on Biogeochemical Carbon and Nitrogen Cycles

4.1 Effect of Microplastics on Carbon Cycles

The ocean is the world's largest repository of activated carbon and plays a vital role in global climate change. However, the release of large amounts of plastic particles will not only damage the marine ecology, but also affect the carbon cycle system of the marine environment.

Phytoplankton and zooplankton are the major producers and consumers in the marine environment. And previous researches have evidenced that microplastics could affect phytoplankton and zooplankton in various ways, which would ultimately influence the marine carbon stocks. As is known to all, the ocean is an important gathering place of CO_2, and carbon sequestration in the ocean has a significant effect in reducing global warming and the greenhouse effect. The global carbon cycle will also undergo great fluctuations when the ocean's ability to absorb carbon dioxide is affected [97]. The primary productivity of marine accounts for about 80% of the planet's total oxygen production. The organic matter and O_2 can be produced by phytoplankton using CO_2 for photosynthesis. They are the main producers in the marine ecosystem. Some studies indicated that microplastics will inhibit the growth of phytoplankton, change the structure of the phytoplankton community, thereby reducing the intensity of photosynthesis, and ultimately disrupting the ocean carbon cycles [78, 80].

In fact, microplastics affect not only phytoplankton but also zooplankton. As the first and foremost important consumer of phytoplankton, zooplankton plays a vital role in the regeneration of marine nutrients, the recycling of biological elements, the transfer of energy, and the transmission of genetic information through the food web. The organic carbon in the ocean can be consumed by zooplankton, which would indirectly influence the remineralization of organic carbon and the ocean-atmosphere carbon dioxide cycle.

Microplastics in the ocean can affect zooplankton and balance the ocean's carbon cycle. Phytoplankton has a certain carbon sequestration ability, but the exposure of microplastics will lead to a reduction in the carbon sequestration ability of phytoplankton and a decrease in the feeding ability of zooplankton. Cole et al. [98] have shown that the microplastics would not only increase the mortality of zooplankton (copepods), but also cause zooplankton to feel full, thereby reducing absorption and consumption of carbon [98]. Over time, the organic carbon consumed by zooplankton may decrease dramatically as the concentration of microplastics increases.

The ubiquitous distribution of microplastics in the ocean has been demonstrated clearly and may influence the ecological balance in the marine food chain/net, thereby interfering with the sea-air exchange and organic carbon processes [97]. The microbial carbon dioxide pump and the micro-biological carbon pump are the main methods for the marine sequestration of carbon dioxide. The former refers to the process of transferring carbon elements from the surface of the ocean to the deep layer in the marine ecological environment, which is driven by biology or biological behavior. Phytoplankton converts inorganic carbon into particulate organic carbon through photosynthesis, self-deposition, and feeding by zooplankton, and finally transfers it to the area of the deep sea. The latter is used by microorganisms to convert active dissolved organic carbon into refractory organic carbon to increase its residence time in the ocean [97].

The zooplankton converts the ingested phytoplankton into feces, through which microplastics are transported to the deep ocean, where the microplastic-containing feces eventually end up in seafloor sediments. However, some studies had pointed out that when microplastics are contained in fecal particles, the sinking rate of fecal particles will slow down and be more prone to rupture, which is not conducive to the deposition to the floor of the deep sea. And this shall reduce the ability of carbon sequestration of the ocean and damage the carbon cycle balance of the ocean.

The existing studies had shown that the presence of microplastics can seriously impact the carbon sequestration capacity of the ocean. As the number of microplastics entering the marine environment increases, its impact on the marine carbon cycle will accelerate as well. Therefore, it is urgently needed to study the effects of microplastics on the marine carbon cycle and to further investigate the potential mechanisms and the scale of such effects.

4.2 Effect of Microplastics on Nitrogen Cycles in Sediments

The ocean nitrogen cycle is a very important link in the earth's elemental cycle. The marine nitrogen cycle is a biochemical process consisting of a series of redox reactions. Nitrogen fixation and nitrogen assimilation provide biological nitrogen (ammonium salt) for the ecosystem. Ammonium salts form nitrates by nitrification and can be converted into nitrogen by denitrification. The whole nitrogen cycle realizes the conversion between different nitrogen-bearing inorganic salts in the ocean. Extensive studies had shown that microplastics in the ocean can either directly or indirectly affect the nitrogen cycle. Microplastics indirectly affect the nitrogen cycle by changing the abundance of biomes and microbiomes in sediments. Marine sediments at the bottom of the microorganisms are of great significance to the marine nitrogen cycle.

In the process of microbial mediated catabolism, nitrification and denitrification have an indispensable link in the nitrogen cycle, nitrification provides the material basis for denitrification, while denitrification can reduce the NO_3^- of the ocean which is helpful for removing the reactive nitrogen in the marine environment to decline the toxicity for nitrate accumulation of marine life.

Since the coastal salt marsh environment is located at the demarcation line between coastal and human-residential areas, it can flow directly into the ocean through land runoff, rainwater discharge, sewage treatment plant discharge, and export. And its vegetation is conducive to the deposition of suspended solids, organic matter, and microplastic particles, so coastal salt marshes are the important area of geochemical cycling [96]. Studies have manifested that the presence of microplastics changes the composition of the microbial community and the nitrogen cycle process in the sediment, and different microplastics have a significant impact on the nitrogen cycle [96]. For instance, polyurethane foam and polylactic acid plastics can promote the nitrification and denitrification processes, while polyvinyl chloride suppresses the process.

Sulfides have been reported to suppress nitrification in marine sediments [99]. Studies have shown that the abundance of sulfate-reducing bacteria in the PVC treatment process is high, and the sulfides produced by the sulfate-reducing bacteria may inhibit the nitrification process [96], thus destroying the nitrogen cycle in the marine environment.

Most of the existing researches only consider the influence of different kinds of microplastics, but seldom include the influence of additives added in microplastics on the nitrogen cycle. Polyvinyl chloride products containing plasticizers are often used in the medical field [100] and have antibacterial properties that affect microbial communities in sediments. The study of Cluzard et al. [101] shows that when PE particles are added to the sediment, the content of NH_4^+ in the overlying water will increase, thus affecting denitrification [101].

Microplastics can affect the nitrogen cycle not only by changing the microbial community richness in the sediment, but also by influencing the regulation of nutrient flux and nitrogen fixation through micro-benthic organisms living at the sediment-water interface [96, 101].

Microplastics regulate marine macrobenthos bioactivity and maintain geochemical cycles. Microbenthos mainly consist of bacteria, cyanobacteria, diatoms, flagellates, amoebas, and ciliates. Microbenthos mainly live in the upper sediments (0–2 mm), and are the main source of many sediment eaters, including, Pratt et al. [102]. Their functions include the decomposition of organic detrital, primary production, and the transfer of substances and energy between trophic levels. It is an important part of the benthic micro-food network and functions as a dispensable role in the benthic ecosystem. Although marine benthos is the main producer, it also relies on organic nitrogen transported to the sediment surface. Marine benthos are a major source of unstable organic matter in soft sediments [103], and studies have suggested that the nitrogen cycle can be affected by both changes in the quality and quantity of organic matter [104]. Marine benthos, therefore, which lives in the sediment-water interface, is also involved in the regulation of nutrients and nitrogen fixation[96].

The net NH_4^+ flux is also related to the feeding, biological activity, and excretion of marine organisms such as Liliana [105], which exists on the surface of sediments and lives by ingestion of marine benthos. The presence of Liliana can increase the efflux of NH_4^+, nevertheless, when the quantity of Liliana is reduced or its functional effect is impaired, it will lead to the decrease of efflux of $NH4^+$ and will inhibit the denitrification process [105, 106]. You et al. reported that the increase of polyethylene terephthalate could affect the physiological reaction of Iliana, and the choice of feeding particles, feeding rate, or physiological activity [107]. Moreover, changes in Liliana' s activities would affect the primary production and nitrogen cycle in sediments [107]. In a sub-nitrogen environment, marine benthos tends to retain and compete for limited nitrogen in the ocean, thereby suppressing the anti-nitrification reaction and resulting in a decrease in denitrification rates.

Microplastics may also interact with benthic organisms to influence denitrification. Huang et al. found that the existence of microplastics or chironomids could promote the process of denitrification and ammoxidation [108]. However, microplastics have a negative effect on benthic invertebrates, causing physiological toxicity. When microplastics and benthic invertebrates coexist, especially when microplastics are exposed at a high concentration and under the positive regulation of Chironomus, the negative nitrogen removal function caused by physiological toxicity will be offset by the positive regulation of Chironomes.

Microplastics not only have a negative effect on the abundance of microbial communities, but also have a positive effect. Studies have shown that microplastics in the sedimentary environment can be used as the organic matter matrix in the microbial environment [96], so as to promote the growth of microorganisms in the sediments and provide a material basis for the development of benthic organisms. Therefore, the interaction between microplastics and benthos is complex.

Currently, the study involves the effect of microplastics and benthic organisms on the nitrogen cycle is still limited. Therefore, it is of great importance to investigate the nitrogen cycle and its mechanism under the joint action of microplastics and benthic organisms.

5 Conclusion and Perspectives

As a new type of pollutant, microplastics have a huge impact on marine ecosystems, which has attracted worldwide attention. Due to the comprehensive influence of physical, chemical, and biological effects, microplastics are widely distributed in various kinds of marine water bodies which will eventually migrate into marine sediments through physical and biological means. The microplastics accumulated in the sediment environment can indirectly influence the carbon and nitrogen cycle in the marine environment by affecting the composition and function of benthic organisms and microbial communities, thereby changing the geochemical cycle process on earth. This article mainly presents the migration, transformation, and fate of microplastics in different marine environments while discussing their impact on plankton, benthic organisms, and microbial communities. Finally, the article summarizes the potential impact of microplastics on the carbon/nitrogen cycle and the mechanisms involved.

The microplastic-associated research deserves more attention, so further studies need to be carried out to explore the following aspects including but not limited to:

(1) The toxicological effects of microplastics on marine organisms at different trophic levels. Microplastics in the marine environment will not only exist in the form of a single pollutant but also release harmful additives. They also tend to absorb hydrophobic organic pollutants as well as heavy metals, thereby generating complicated toxicological effects. Hence, it is of great significance to investigate the toxicological effects of microplastics from different aspects.

(2) The study about the combined effect of microplastics and benthic organisms on geochemical cycles is rarely documented. Therefore, it is worthwhile to study the geochemical cycle and its mechanism under the joint action of microplastics and benthic organisms.

(3) The microplastics that exist in the marine ecological system have a great influence on marine carbon sequestration. The influence of microplastics on ocean carbon sequestration is yet an emerging hot issue and many conclusions are still in the speculation period, thus it is necessary to focus on the potential mechanism and the scale and scope of such impact, especially in the deep sea and seabed sediment environment.

References

1. Xue W, Huang D, Zeng G, Wan J, Zhang C, Rui X, Cheng M (2017) Nanoscale zero-valent iron coated with rhamnolipid as an effective stabilizer for immobilization of Cd and Pb in river sediments. J Hazard Mater 341:381–389
2. PlasticEurope (2019) An analysis of European plastics production, demand and waste data
3. Lebreton L, Van der Zwet J, Damsteeg J-W, Slat B, Andrady AL, Reisser J (2017) River plastic emissions to the world's oceans. Nat Commun 8:15611
4. Thompson RC (2004) Lost at sea: where is all the plastic?. Sci 304(5672):838

5. Ryan PG, Moore CJ, van Franeker JA, Moloney CL (2009) Monitoring the abundance of plastic debris in the marine environment. Philos Trans Royal Soc London. Ser B, Biol Sci 364(1526):1999–2012
6. Andrady AL (2011) Microplastics in the marine environment. Mar Pollut Bull 62(8):1596–1605
7. Hidalgo-Ruz V, Gutow L, Thompson RC, Thiel M (2012) Microplastics in the marine environment: a review of the methods used for identification and quantification. Environ Sci Technol 46(6):3060–3075
8. Li J, Huang W, Jiang, R, Han X, Zhang D, Zhang C (2020) Are bacterial communities associated with microplastics influenced by marine habitats? Sci Total Environ 733:139400
9. Zhang D, Liu X, Huang W, Li J, Wang C, Zhang D, Zhang C (2020) Microplastic pollution in deep-sea sediments and organisms of the Western Pacific Ocean. Environ Pollut 259:113948
10. Zhang D, Cui Y, Zhou H, Jin C, Yu X, Xu Y, Li Y, Zhang C (2019) Microplastic pollution in water, sediment, and fish from artificial reefs around the Ma'an Archipelago, Shengsi, China. Sci Total Environ 703:134768
11. Cole M, Lindeque P, Halsband C, Galloway TS (2011) Microplastics as contaminants in the marine environment: a review. Mar Pollut Bull 62(12):2588–2597
12. Zhang K, Shi H, Peng J, Wang Y, Xiong W (2018) Microplastic pollution in China's inland water systems: a review of findings, methods, characteristics, effects, and management. Sci Total Environ 630:1641–1653
13. Alomar C, Estarellas F, Deudero S (2016) Microplastics in the Mediterranean sea: deposition in coastal shallow sediments, spatial variation and preferential grain size. Mar Environ Res 115:1–10
14. Li J, Huang W, Yongjiu X, Jin A, Zhang D, Zhang C (2020) Microplastics in sediment cores as indicators of temporal trends in microplastic pollution in Andong salt marsh, Hangzhou Bay, China. Reg Stud Mar Sci 35:101149
15. WoodallL, Sanchez-Vidal A, Canals M, Gordon L, Paterson J, Coppock R, Sleight V, Calafat A, Rogers AD, Narayanaswamy BE, Thompson RC (2014) The deep sea is a major sink for microplastic debris. Royal Soc Open Sci 1(4):140317
16. Evangeliou N, Grythe H, Klimont Z, Heyes C, Stohl A (2020) Atmospheric transport is a major pathway of microplastics to remote regions. Nat Commun 11(1):3381
17. Radisic V, Nimje PS, Bienfait AM, Nachiket P (2020) Marine plastics from norwegian west coast carry potentially virulent fish pathogens and opportunistic human pathogens harboring new variants of antibiotic resistance genes. Microorganisms 8(8):1200
18. Bakir A, Rowland SJ, Thompson RC (2014) Transport of persistent organic pollutants by microplastics in estuarine conditions. Estuarine Coastal Shelf Sci 140:14–21
19. Kusum KK, Vineetha G, Raveendran TV, Muraleedharan KR, Nair M, Achuthankutty CT (2011) Impact of oxygen-depleted water on the vertical distribution of chaetognaths in the northeastern Arabian Sea. Deep Sea Res Part I 58(12):1163–1174
20. Bowley J, Baker-Austin C, Hartnell R, Lewis C, Porter A (2020) Oceanic hitchhikers—assessing pathogen risks from marine microplastic. Trends Microbiol 29(2):107–116
21. Carbery M, O'Connor W, Thavamani P (2018) Trophic transfer of microplastics and mixed contaminants in the marine food web and implications for human health. Environ Int 115(JUN):400–409
22. Koelmans AA, Besseling E, Foekema EM (2014) Leaching of plastic additives to marine organisms. Environ Pollut 187:49–54
23. Liorca M, Abalos M, Vega-Herrera A, Adrados MA, Abad E, Farre M (2020) Adsorption and desorption behaviour of polychlorinated biphenyls onto microplastics' surfaces in water/sediment systems. Toxics 8(3):59
24. Marris E (2014) Fate of ocean plastic remains a mystery. Nature News. https://doi.org/10.1038/nature.2014.16508
25. Eriksen M, Laurent LCM, Carson HS, Thiel M, Moore CJ, Borerro JC, Galgani F, Ryan PG, Reisser J (2014) Plastic pollution in the world's oceans: more than 5 trillion plastic pieces weighing over 250,000 tons Afloat at Sea. Plos One 9(12):e111913

26. Francesca MCF, Ryan PG (2016) Biofouling on buoyant marine plastics: an experimental study into the effect of size on surface longevity. Environ Pollut 210:354–360
27. Song Y, Hong S, Jang M, Han G, Shim W (2015) Occurrence and distribution of microplastics in the sea surface microlayer in jinhae bay, south korea. Arch Environ Con Tox 69(3):279–287
28. Zhao S, Zhu L, Wang T, Li D (2014) Suspended microplastics in the surface water of the Yangtze Estuary System, China: first observations on occurrence, distribution. Mar Pollut Bull 86:562–568
29. Isobe A, Uchiyama-Matsumoto K, Uchida K, Tokai T (2016) Microplastics in the Southern Ocean. Mar Pollut Bull 114:623–626
30. Nel HA, Hean JW, Noundou XS, Froneman PW (2017) Do microplastic loads reflect the population demographics along the southern african coastline?. Mar Pollut Bull 115(1–2):115–119
31. Nel HA, Froneman PW (2015) A Quantitative Analysis of Microplastic Pollution Along the South-eastern Coastline of South Africa. Mar Pollut Bull 101:274–279
32. Sutton R, Mason SA, Stanek SK, Willis-Norton E, Wren IF, Box C (2016) Microplastic contamination in the san francisco bay, california, usa. Mar Pollut Bull109(1):230–235
33. Mauro RD, Kupchik MJ, Benfield MC (2017) Abundant plankton-sized microplastic particles in shelf waters of the northern Gulf of Mexico. Environ Pollut 230:798–809
34. Suaria G, Avio CG, Lattin G, Regoli F, Aliani S (2017) Floating microplastic in the South Adriatic sea. Fate and impact of microplastics in marine ecosystems. From the coastline to the open sea, 51–52
35. De Carvalho DG, Baptista Neto JA (2016) Microplastic pollution of the beaches of Guanabara Bay, Southeast Brazil. Ocean Coastal Manag 128:10–17
36. Obbard RW, Sadri S, Wong YQ, Khitun AA, Baker I, Thompson RC (2014) Global warming releases microplastic legacy frozen in Arctic Sea ice. Earths Fut 2(6):315–320
37. Cole M, Lindeque P, Fileman E, Halsband C, Goodhead R, Moger J, Galloway TS (2013) Microplastic Ingestion by Zooplankton. Environ Sci Technol 47(12):6646–6655
38. Maes T, Van der Meulen MD, Devriese LI, Leslie HA, Frere L, Robbens J, Dick Vethaak A (2017) Microplastics baseline surveys at the water surface and in sediments of the North-East Atlantic. Front Mar Sci 4:135
39. Zhang C, Zhou H, Cui Y, Wang C, Li Y, Zhang D (2018) Microplastics in offshore sediment in the Yellow Sea and East China Sea, China. Environ Pollut 244:827–833
40. Kunz A, Walther BA, Lowemark L, Lee Y-C (2018) Distribution and quantity of microplastic on sandy beaches along the northern coast of Taiwan. Mar Pollut Bull 111(1–2):126–135
41. Kazmiruk TN, Kazmiruk VD, Bendell L (2018) Abundance and distribution of microplastics within surface sediments of a key shellfish growing region of Canada. Plos One 13(5):e0196005
42. Abidli S, El Menif NT, Toumi H, Lahbib Y (2017) The first evaluation of microplastics in sediments from the complex lagoon-channel of Bizerte (Northern Tunisia). Water Air Soil Pollut 228(7):262
43. Ballent A, Corcoran PL, Madden O, Helm PA, Longstaffe FJ (2016) Sources and sinks of microplastics in Canadian Lake Ontario nearshore, tributary and beach sediments. Mar Pollut Bull 110(1):383–395
44. Satoshi N, Asako O, Kazuo Y, Keiko M, Tadashi N, Takanori S (2019) Microplastics contamination in tidelands of the Osaka bay area in western Japan. Water Environ J 34(3):474–488
45. Manalu AA, Hariyadi S, Wardiatno Y (2017) Microplastics abundance in coastal sediments of Jakarta Bay, Indonesia. AACL Bioflux 10(5):1164–1173
46. Van Cauwenberghe L, Claessens M, Vandegehuchte, MB (2015) Microplastics are taken up by mussels (Mytilus edulis) and lugworms (Arenicola marina) living in natural habitats. Environ Pollution 199:10–17
47. Kolandhasamy P, Lei S, Li J, Qu X, Jabeen K, Shi H (2018) Adherence of microplastics to soft tissue of mussels: a novel way to uptake microplastics beyond ingestion. Sci Total Environ 610–611:635–640

48. Santana MFM, Ascer LG, Custodio MR, Moreira FT, Turra AJMPB (2016) Microplastic contamination in natural mussel beds from a Brazilian urbanized coastal region: Rapid evaluation through bioassessment. Mar Pollut Bull 106(1-2):183–189
49. Renzi M, Guerranti C, Blaslovic AJMPB (2018) Microplastic contents from maricultured and natural mussels. Mar Pollut Bull 131:248–251
50. Khan MB, Prezant, Prezent RS (2018) Microplastic abundances in a mussel bed and ingestion by the ribbed marsh mussel Geukensia demissa. Mar Pollut Bull 130:67–75
51. Catarino AI, Macchia V, Sanderson WG, Thompson RC, Henry TBJEP (2018) Low levels of microplastics (MP) in wild mussels indicate that MP ingestion by humans is minimal compared to exposure via household fibres fallout during a meal. Environ Pollut 237:675–684
52. Frias JPGL, Otero V, Sobral P (2014) Evidence of microplastics in samples of zooplankton from Portuguese coastal waters. Mar Environ Res 95:89–95
53. Li Q, Zhu M, Liang J, Zheng S, Zhao Y (2017) Ingestion of microplastics by natural zooplankton groups in the northern South China Sea. Mar Pollut Bull 115(1–2):217–224
54. Neves D, Sobral P, Ferreira JL, Pereira T (2015) Ingestion of microplastics by commercial fish off the Portuguese coast. Mar Pollut Bull 101:119–126
55. Alomar C, Estarellas F, Deudero S (2016) Microplastics in the Mediterranean sea: deposition in coastal shallow sediments, spatial variation and preferential grain size. Mar Environ Res 115:1–10
56. Abbasi S, Soltani N, Keshavarzi B, Moore F, Turner A, Hassanaghaei M (2018) Microplasticsin different tissues of fish and prawn from the Musa Estuary, Persian Gulf. Chemosphere 205:80–87
57. Jabeen K, Su L, Li J, Yang D, Tong C, Mu J (2016) Microplastics and mesoplastics in fish from coastal and fresh waters of china. Environ Pollut 221:141–149
58. Bessa F, Barría P, Neto JM, Frias JPGL, Otero V, Sobral P (2018) Occurrence of microplastics in commercial fish from a natural estuarine environment. Mar Pollut Bull 128:575–584
59. Fadiyah M, Baalkhuyur, El-Jawaher A (2018) Microplastic in the gastrointestinal tract of fishes along the saudi arabian red sea coast. Mar Pollut Bull 131:407–415
60. Vendel AL, Bessa F, Alves VEN, Amorim ALA, Patrício J, Palma ART (2017) Widespread microplastic ingestion by fish assemblages in tropical estuaries subjected to anthropogenic pressures. Mar Pollut Bull 117(1–2):448–455
61. Steer M, Lindeque PK, Cole M, Thompson R (2018) Microplastic ingestion in fish larvae in the western english channel. Environ Pollut 1–10
62. Zhao Y, Sun X, Li Q, Shi Y, Zheng S, Liang J (2019) Data on microplastics in the digestive tracts of 19 fish species from the yellow sea, china. Data in brief 25:103989
63. Abadi ZTR, Abtahi B, Grossart HP, Khodabandeh S (2021) Microplastic content of kutum fish, rutilus frisii kutum in the southern caspian sea. Sci Total Environ 752(15):141542
64. Borrelle SB, Avery-Gomm S, Provencher J (2016) Room for Improvement: spatial, taxonomic and methodological gaps in seabird plastic ingestion research. Society for conservation biology oceania. https://doi.org/10.13140/RG.2.2.22020.12160
65. Amélineau, Bonnet, Heitz, Mortreux AAM, Harding (2016) Microplastic pollution in the greenland sea: background levels and selective contamination of planktivorous diving seabirds. Environ Pollut 219:1131–1139
66. Terepocki AK, Brush AT, Kleine LU, Shugart GW, Hodum P (2017) Size and dynamics of microplastic in gastrointestinal tracts of Northern Fulmars (Fulmarus glacialis) and Sooty Shearwaters (Ardenna grisea). Mar Pollut Bull 116(1–2):143–150
67. Caron AGM, Thomas CR, Berry KLE, Motti CA, Ariel E, Brodie JE (2018) Ingestion of microplastic debris by green sea turtles (chelonia mydas) in the great barrier reef: validation of a sequential extraction protocol. Mar Pollut Bull 127:743–751
68. Nava V, Leoni B (2021) A critical review of interactions between microplastics, microalgae and aquatic ecosystem function. Water Res 188:116476
69. Michels J, Stippkugel A, Wirtz K, Engel A (2015) Aggregation of microplastics with marine biogenic particles. Aslo Aquatic Sciences Meeting 2015

70. Kalcikova G, Skalar T, Marolt G, Kokalj AJ (2020) An environmental concentration of aged microplastics with adsorbed silver significantly affects aquatic organisms. Water Res175:115644.1–115644.9
71. Underwood GJC, Boulcott M, Raines CA, Waldron K (2010) Environmental effects on exopolymer production by marine benthic diatoms: dynamics, changes in composition, and pathways of production. J Phycol 40(2):293–304
72. Wirnkor A, EbereEC, Ngozi E, Oharley N (2019) Microplastic–toxic chemical interaction: a review study on quantified levels, mechanism and implication. SN Appl Sci 1(11):1400
73. Yeung WYK (2017) Ecological impacts of larvicidal oil on the marine ecosystem: implications on its management. The University of Hong Kong
74. Long M, Paul-Pnot I, Hégaret H, Moriceau B, Lambert C, Huvet A, Soudant P (2017) Interactions between polystyrene microplastics and marine phytoplankton lead to species-specific hetero-aggregation. Environ Pollution 228:454–463
75. Priyanka B, Sijie L, James P, Turner P, Chun K (2010) Physical adsorption of charged plastic nanoparticles affects algal photosynthesis. J Phys Chem 114(39):16556–16561
76. Chae Y, Kim D, An Y-J (2019) Effects of micro-sized polyethylene spheres on the marine microalga Dunaliella salina: Focusing on the algal cell to plastic particle size ratio. Aquatic Toxicol (Amsterdam, Netherlands) 216:105296
77. Wang C, Dong X, Shao Q, Pan X, Qin P (2019) Research progress on pollution behavior and toxicological effects of nanoplastics. Guangdong Chem Ind 46(9):138–139
78. Zhang C, Chen X, Wang J, Tan L (2016) Toxic effects of microplastic on marine microalgae Skeletonema costatum: interactions between microplastic and algae. Environ Pollut 220:1282–1288
79. Mao Y, Ai H, Zhang Z, Zeng P, Kang L, Li W, Weikang G, He Q, Li H (2018) Phytoplankton response to polystyrene microplastics: perspective from an entire growth period. Chemosphere 208:59–68
80. Ting Z, Liju T, Wenqiu H, Jiang W (2019) The interactions between micro polyvinyl chloride (mPVC) and marine dinoflagellate Karenia mikimotoi: The inhibition of growth, chlorophyll and photosynthetic efficiency. Environ Pollut 247:883–889
81. Besseling E, Wang B, Luerling M, Koelmans AA (2014) Nanoplastic affects growth of S. obliquus and reproduction of D. magna. Environ Sci Technol 48(20):12336–12343
82. Kosore CM, Ojwang L, Maghanga JK, Kamau JN, Kimeli A, Omukoto J, Ngisiag' N, Mwaluma J, Ong'ada H, Magori C (2018) Occurrence and ingestion of microplastics by zooplankton in Kenya's marine environment: first documented evidence. African J Mar Sci 40(3):225–234
83. Shen M, Ye S, Zeng G, Zhang Y, Xing L, Tang W, Wen X, Liu S (2019) Can microplastics pose a threat to ocean carbon sequestration? Mar Pollut Bull 150:110712
84. Wieczorek AM, Croot PL, Lombard F, Sheahan JN, Doyle TK (2019) Microplastic ingestion by gelatinous zooplankton may lower efficiency of the biological pump. Environ Sci Technol 53(9):5387–5395
85. Volkenborn N, Hedtkamp SIC, Beusekom JEEV, Reise KJEC (2007) Effects of bioturbation and bioirrigation by lugworms (Arenicola marina) on physical and chemical sediment properties and implications for intertidal habitat succession. Estuarine Coastal Shelf Sci 74(1–2):331–343
86. Norling K, Rosenberg R, Hulth S, Gremare A, Bonsdorff E (2008) Importance of functional biodiversity and species-specific traits of benthic fauna for ecosystem functions in marine sediment. Mar Ecol Progr 332:11–23
87. Wegner A, Besseling E, Foekema EM, Kamermans P, Koelmans AA (2012) Effects of nanopolystyrene on the feeding behavior of the blue mussel (Mytilus edulis L.). Environ Toxicol Chem 31(11):2490–2497
88. Rist S, Baun A, Almenda R, Hartmann NB (2019) Ingestion and effects of micro- and nanoplastics in blue mussel (Mytilus edulis) larvae. Mar Pollut Bull 207:423–430
89. Evan Ward J, Zhao S, Holohan BA, Mladinich Mladinich K, Griffin T, Wozniak J, Shumway SE (2019) Selective ingestion and egestion of plastic particles by the blue mussel (Mytilus edulis)

and eastern oyster (Crassostrea virginica): implications for using bivalves as bioindicators of microplastic pollution. Environ Sci Technol 53(15):8776–6784
90. Silva Carlos JM, Patricio Silva AL, Campos D, Soares Amadeu MVM, Pestana Joao LT, Gravato C (2020) Lumbriculus variegatus (oligochaeta) exposed to polyethylene microplastics: biochemical, physiological and reproductive responses. Ecotoxicol Environ Saf 207:111375
91. Li C, Gan Y, Dong J, Fang J, Chen H, Quan Q, Liu J (2020) Impact of microplastics on microbial community in sediments of the Huangjinxia Reservoir—water source of a water diversion project in western China, China. Chemosphere 253:126740
92. Mccormick A, Hoellein TJ, Mason SA, Schluep J, Kelly JJ (2014) Microplastic is an abundant and distinct microbial habitat in an urban river. Environ Sci Technol 48(20):11863–11871
93. Harrison JP, Schratzberger M, Sapp M, Osborn AM (2014) Rapid bacterial colonization of low-density polyethylene microplastics in coastal sediment microcosms. BMC Microbiol 14:232
94. Miao L, Wang P, Jun Hou Yu, Yao ZL, Liu S, Li T (2018) Distinct community structure and microbial functions of biofilms colonizing microplastics. Sci Total Environ 650:2395–2402
95. Kooi M, van Nes EH, Scheffer M, Koelmans AA (2017) Ups and downs in the ocean: effects of biofouling on vertical transport of microplastics. Environ Sci Technol 51(14):7963–7971
96. Seeley ME, Song B, Passie R, Hale RC (2020) Microplastics affect sedimentary microbial communities and nitrogen cycling. Nat Commun 11(1):2372
97. Shen M, Zhu Y, Zhang Y, Zeng G, Wen X, Yi H, Ye S, Ren X, Song B (2019) Micro(nano)plastics: unignorable vectors for organisms. Mar Pollut Bull 139:328–331
98. Cole M, Lindeque P, Fileman E, Halsband C, Galloway TS (2015) The impact of polystyrene microplastics on feeding, function and fecundity in the marine copepod Calanus helgolandicus. Environ Sci Technol 49(2):1130–1137
99. Ward BB (2008) Nitrification in Marine systems. Nitrogen in the Marine environment (2nd Edition), pp 199–261
100. Choi SY, Rodríguez H, Nimal Gunaratne HQ, Puga AV, Gilpin DF, Mcgrath S, Vyle JS, Tunney MM, Rogers RD, Mcnally T (2014) Dual functional ionic liquids as antimicrobials and plasticisers for medical grade PVCs. RSC Adv 4(17):8567–8581
101. Cluzard M, Kazmiruk TN, Kazmiruk VD, Bendell L (2015) Intertidal concentrations of microplastics and their influence on ammonium cycling as related to the shellfish industry. Arch Environ Contamination Toxicol 69(3):310–319
102. Pratt DR, Pilditch CA, Lohrer AM, Thrush SF, Kraan C (2015) Spatial distributions of grazing activity and microphytobenthos reveal scale-dependent relationships across a sedimentary gradient. Estuaries Coasts 38(3):722–734
103. Tobias C, Giblin A, McClelland J, Tucker J, Peterson B (2003) Sediment DIN fluxes and preferential recycling of benthic microalgal nitrogen in a shallow macrotidal estuary. Mar Ecol Progr 257(8):25–36
104. Fulweiler RW, Brown SM, Nixon SW, Jenkins BD (2013) Evidence and a conceptual model for the co-occurrence of nitrogen fixation and denitrification in heterotrophic marine sediments. Mar Ecol Prog Ser 482:57–68
105. Woodin S, Volkenborn N, Pilditch CA, Lohrer AM, Wethey DS, Hewitt JE, Thrush SF (2016) Same pattern, different mechanism: locking onto the role of key species in seafloor ecosystem process. Sci Rep 6:26678
106. Gladstone-Gallagher RV, Hughes RW, Douglas EJ, Pilditch CA (2018) Biomass-dependent seagrass resilience to sediment eutrophication. J Exp Mar Biol Ecol 501:54–64
107. You Y, Thrush SF, Hope JA (2020) The impacts of polyethylene terephthalate microplastics (mPETs) on ecosystem functionality in marine sediment. Mar Pollut Bull 160:111624
108. Huang Y, Li W, Wang F, Yao J, Yang W, Han L, Lin D, Min B, Zhi Y, Grieger K, Yao J (2021) Effect of microplastics on ecosystem functioning: microbial nitrogen removal mediated by benthic invertebrates. Sci Total Environ 754:142133

Domestic Laundry and Microfiber Shedding of Synthetic Textiles

R. Rathinamoorthy and S. Raja Balasaraswathi

Abstract Microplastic pollution is one of the recent issues that attracted many environmental researchers. Many in situ samplings reported that more than 80% of the microplastics found in the ocean and sea bed are microfibers from textiles. Microfiber shedding is the phenomenon of disengagement of loose or damaged fibers from the surface of the textile materials. The textile materials can be subjected to stress and abrasion during all the stages of the life cycle starting from manufacturing to disposal. Among all phases, during the usage phase, the laundry process causes maximum damage of up to 90% to the clothing and this is the reason why the durability of the garments is measured based on the number of washes it can withstand (Carr in Chemistry of textiles industry. Blackie Academic and Professional, London [1]). Hence, the laundering of synthetic textile is identified as one of the major sources of microfiber shedding. Another reason for domestic laundry to play an important role in microfiber pollution is that they can directly pollute the water bodies as wastewater. This chapter details the importance of the issue along with the effect of laundry on shedding. The chapter further elucidates the different influencing parameters like washing machine type, washing temperature, washing duration, the addition of detergents, usage of laundry additives on microfiber shedding. The potential control measures of the shedding are also discussed to avoid microfiber pollution through the laundry.

Keywords Microfiber shedding · Synthetic textile · Domestic laundry · Aging of garment · Laundry machine type · Detergent · Softeners · Laundry temperature · Laundry time

---

R. Rathinamoorthy (✉) · S. Raja Balasaraswathi
Department of Fashion Technology, PSG College of Technology, Coimbatore, India
e-mail: r.rathinamoorthy@gmail.com

S. S. Muthu (ed.), *Microplastic Pollution*, Sustainable Textiles: Production, Processing, Manufacturing & Chemistry, https://doi.org/10.1007/978-981-16-0297-9_5

1 Introduction

Microfibers are a type of microplastics, which are generally released from the synthetic textile material, whereas microplastics are common plastic particles that have other non-textile sources. Microfiber is a synthetic textile fiber with a diameter of fewer than 10 micrometers and with a higher length to diameter ratio [2]. Recent research data from environmental investigators reported that microfibers are one of the rising pollutions to the aquatic system. Mishra et al. reported that approximately 1.5 million trillion microfibers were already deposited in the ocean beds and every year around 2 million tons are released into the ocean [3]. Though synthetic textile materials are referred to as microfibers that contribute to microplastics, any fibers in <10 μm are considered to be microfiber, irrespective of the origin (natural or synthetic). Studies also reported that in some regions, the microfibers present in the freshwater system and air were dominated by natural fiber too. However, compared to the synthetic textiles, the biodegradability of the natural textile materials made their impact on the environment as an insignificant one [4]. Several research results indicated that the majority of the microplastics found were microfiber types in several parts of the globe. Many in situ sampling studies reported the presence of microfiber as a major pollutant in the sea and ocean bed. In a detailed study, the researchers have collected samples from 18 different seashores across six continents and measured for microplastic pollution. The results showed that approximately 2–31 microplastics were found in every 250 ml of the water sample. Out of which, polyester (56%) occupies the majority of the plastics found followed by acrylic (23%), polypropylene (7%), polyethylene (6%), and polyamide fibers (3%). The results also showed a strong positive correlation between the human population density and microplastic concentration. Out of all the fiber types, polyester and acrylic fibers were dominating in every sample. Hence, in their study, they have evaluated and confirmed that the microfibers are the most common pollutant type and those were mainly derived from sewage due to the cloth laundry rather than other modes of microplastic generation namely fragmentation, cleaning of products [5].

A similar study performed in the Swedish coastal area confirmed the presence of microfibers in the ocean beds. The research identified that the synthetic textile, tyre abrasives, and fishnets are major sources of microfiber pollution in seawater. Out of which, 90% of the particles obtained were microfibers from the synthetic textiles [6]. The other report estimated the microfiber release based on the assumption from literature as no direct measurement of the microfiber emission is not available. They have assumed that 30–50% of used clothing was made off synthetic textiles and the volume of the synthetic textiles laundered per capita per year is 200–300 kg. Based on their predictions, results reported that for the population of 9.85 million (as of 2015), the average mass of the microfibers released per year was 8–945 tons. The results also reported that approximately 12–640 mg microfiber sheds per kg of synthetic fabric washing [7]. Woodall et al. analyzed the deep-sea samples from different parts of the world and confirmed the microfiber presence. Their results showed that for every 50 ml of water, approximately 1.4–4 pieces of microfibers were found in the samples. From their results, they have estimated that approximately four billion microfibers were present per square kilometer of the Indian Ocean, fifteen billion in the Atlantic and Mediterranean Sea [8]. Though we are using synthetic textile for a longer time, the problem becomes serious in recent times due to the prevailing Fast Fashion trend. Fast fashion is an activity which mimics and provides high fashion garments at cheaper and affordable rate to all range of customers. Fast fashion trends also offer speedy fashion trend changes so that the customers can purchase new and trendy clothing more often. Hence, in order to reduce the cost and frequently deliver the product in the market, the manufacturers use synthetic textiles. Due to this, synthetic fiber consumption in the fashion industry has tremendously increased in the past decade.

A recent report from the Ellen MacArthur Foundation supported the increased use of synthetic textiles in recent times. The research report mentioned that in the past fifteen years the clothing utilization significantly reduced approximately by 36% in lower income and 70% in the higher income countries due to the fast fashion trend. On microfiber estimation, they mentioned that in between 2015 and 2050, 22 million tons of microfibers will be added into the ocean due to the higher use of synthetic textile [9]. According to IHS Markit's Polyester fiber's 2018 market, the annual growth rate of polyester fiber since the year 2000 is 6%. The large industrial application, as a substitution of other material and low-cost nature, are the main factors for its wider growth. The polyester production has doubled from 24.7–53.7 million metric tons in 12 years from 2005 to 2017. Due to the cheap and versatile nature, polyester officially overtook cotton in 2002 and became the number one fiber (higher consumption) in the world [10]. The higher consumption of synthetic textiles increases the higher exposure of synthetic material to the environment. The exposed microplastic undergoes various changes like physical, enzymatic, and microbial degradation and is broken down into a very small size and cannot be degraded further. It is very hard to remove such contamination from soil or water bodies as that will result in the removal of all plankton-size organisms. Microfibers are also noted as airborne pollutants from several clothing manufacturing hubs. Other than environmental impact, microfiber pollution also leads to health hazards to humans in different aspects. Recent research was performed on the drinking water samples from bottled waters of different brands from 19 locations in 9 countries. The results of the study showed an average of 325 plastic particles in every liter of water sold in the bottle. Out of 259 bottles tested, only 17 bottles were noted free from microplastics. Similarly, a survey examined tap water from 5 different continents and reported that 83% of the samples were noted with microplastic pollution. Out of the waters examined, with 82.4% of microplastic contaminations, water from India obtained a rank of 3 for highly polluted water [11, 12]. To threaten human life, a few food items are also found with microfibers and plastics as reported by other researchers including seafood, honey, table salt, beer, and also in fresh vegetables [12–15].

Recent research shocked with the information that the rainwater samples collected and analyzed in the national parks and wilderness areas across Colorado, which is a protected region, had microfiber particles in it. The researchers were also surprised to see plastic fragments <5 mm which are mostly of microfibers. Through 32 different particle scanners, they have concluded that roughly 4% of the analyzed atmospheric particles are synthetic. In particular, most of the particles found in the wet and dry samples are microfibers of bright colors mainly originated from clothing but not from the commercial personal care product [16]. These findings were more threatening, further to add, the researchers reported that the current deposit rate of microplastic is 132 plastics per square meter per day and that adds more than 1000 metric tons of plastics in the western protected lands of the US annually. It is estimated to accumulate approximately 11 billion metric tons of plastics in the environment by 2025. A detailed analysis of the breathtaking disaster can be found in the reference for further knowledge [17].

2 Microfiber Shedding and Laundry

Microfiber shedding from textile material is a mechanical phenomenon, where due to the application of external force, the loose fiber or damaged fiber disengages and sheds from the cloth. Any physical process like manufacturing, handling, usage, and disposal creates mechanical force and motivates the shedding of microfiber. The shedding of microfiber can be directly related to the textile properties of fabric like tensile strength, pill formation, and abrasion resistance nature of the fabrics [18]. The fiber breakage in textile material may happen in two different aspects. Firstly, due to mechanical forces like tensile fatigue and flex fatigue are the two reasons for the fiber damage. Hearle et al. mentioned that during usage of clothing, repeated applications of small force and loading will break the textile material rather than an excessive force. The fiber damage also happens in textile materials due to surface abrasion while wearing and using the textile material [19]. Secondly, the major disengagement of fibers from the structure of the fabric occurs during the washing, in the presence of the chemicals. The washing chemicals like detergent, bleach, and other components chemically react with material during the dirt removal process. Hence at this stage, the weak fiber will get damaged and removed from the surface of the textiles. Irrespective of the type of fibers used, both the natural and synthetic textiles usually shed at every stage of its life. 90% of the mechanical and chemical damages to the clothing occur in the laundry phase and so the maximum microfiber shedding is expected in the laundry phase. In specific, research reports mentioned that a garment can shed approximately 1900–11,000 fibers per wash in a normal domestic washing process. Where the natural fibers are ignored due to its biodegradability but in the case of synthetic textiles, the effect is adverse in multiple areas including environmental and health care impact [20].

The first analysis and estimation of microfiber shedding from synthetic textile material were performed by Brown et al. The researchers initially analyzed the water samples of the ocean at different locations of the globe. As their results pointed more microplastic emission from synthetic textiles, the research performed a laundry trial. On analysis of polyester blanket, fleece, and shirt, they identified that each garment can shed more than 1900 microfibers per laundry. Further, they noted more than

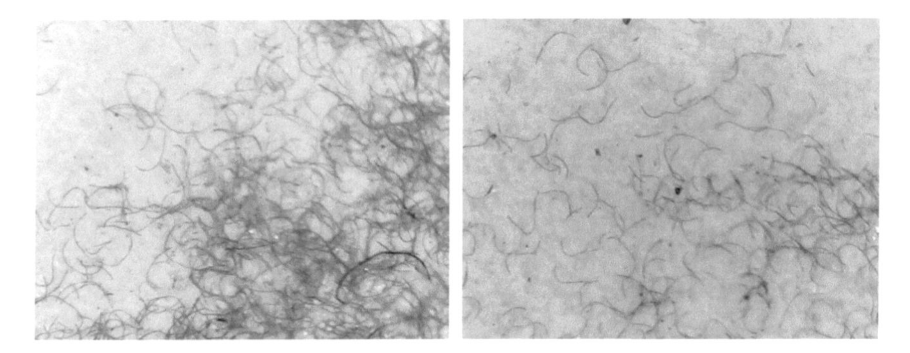

Fig. 1 Microfiber obtained from a polyester knitted fabric after a household laundry (100X, on a filter paper) (Authors own illustration)

100 microfibers per L of the wastewater from the laundry, specifically 180% higher shedding with fleece material. By demonstrating the result, the research concluded laundry as a potential source of microfiber emission [5]. Research results revealed that approximately 35% of the microplastic emission originated from synthetic clothing. Further, it is estimated that 60% of the total global fiber consumption is from synthetic textiles and approximately 69.7 metric tons are used in the apparel industry alone [21]. Figure 1 represents the microfiber fragments obtained from a polyester fabric through domestic laundry.

It is also noted that roughly 840 million domestic washing machines are used globally. The increased usage of synthetic textiles and frequent laundries around the world signals the rising importance in the microfiber shedding through the laundry. Though several measures were taken by governments, washing effluents from various countries still reaches the ocean without proper treatment. A proper sewage treatment system with sand filters and membrane bioreactors can filter 70–99% of the microplastics, but still, more than 1770 microplastics (0.009 particles/L) per hour will reach the aquatic system directly [22] and also to the terrestrial system through the sludge disposal or treatment methods [23].

Other research evaluated the microfiber shedding of 100% polyester, 35–65% pure, recycled polyester, and polyester, cotton, and modal blended fabrics, which are commercially sourced. The researcher laundered the fabric with a commercial detergent and measured the amount of microfiber shedding on a weight basis and also in the count. The results revealed that the shedding ranged from 124 to 308 mg per kg of washed fabric type and the corresponding count represents the number ranging from 640,000 to 1,500,000. Figure 2 represents the average amount of microfibers released after a wash from different types of commercial apparel sourced [24]. The different laundry parameters that have a significant influence on the microfiber shedding from synthetic textiles are provided in Fig. 3.

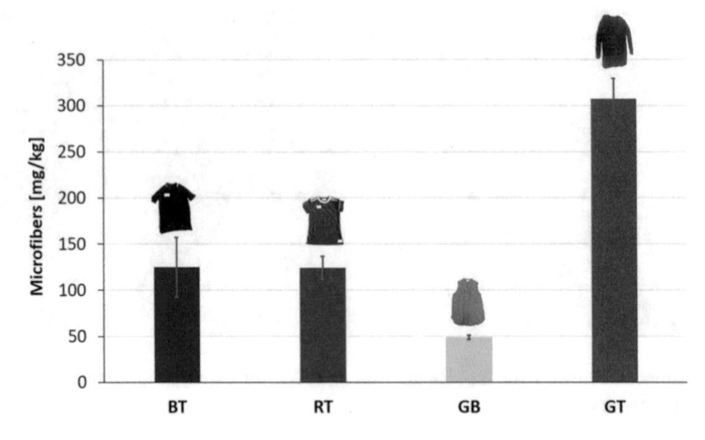

Fig. 2 Microfibers released (expressed in mg/kg, Ma ± SD, $n = 2$) from BT, a 100% polyester t-shirt, RT, a 100% polyester t-shirt, GB, a 100% polyester blouse of which 65% is recycled polyester, and GT, a top whose front is made of 100% polyester and whose back is made of a blend of 50% cotton and 50% modal [24]

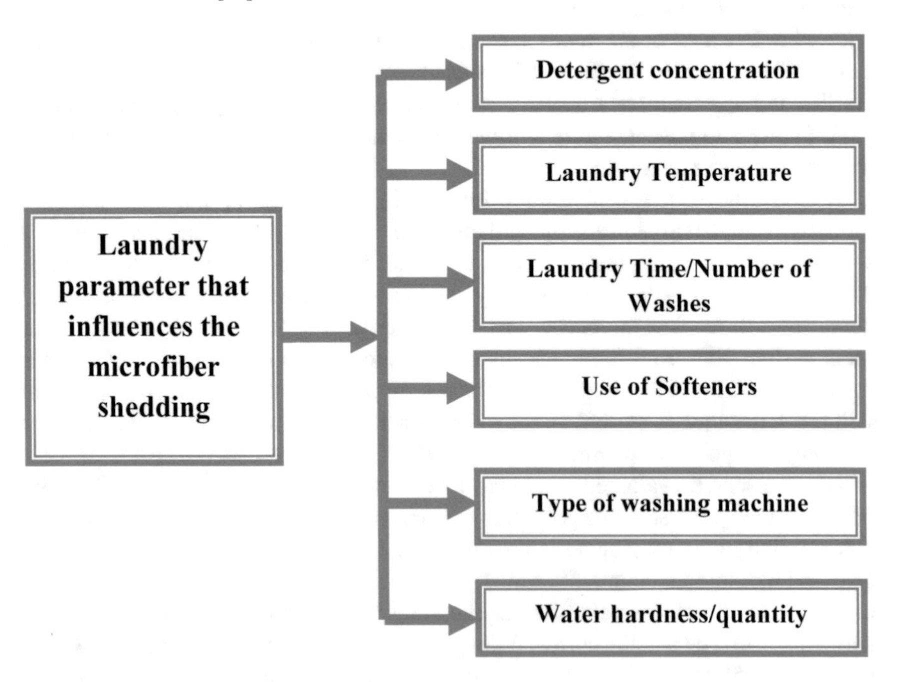

Fig. 3 Laundry parameters influencing the microfiber shedding from textiles (Authors own illustration)

2.1 Effect of Detergent/Detergent Concentration on Microfiber Shedding

Detergents are commercially used in the laundry process to remove soils, dirt, and other stains from the textile surface. The main ingredient in the laundry detergent is surfactants. The surfactant is the substance that has two different ends one easily attaches with water and another end easily binds with dirt in the textiles. The surfactant reduces the surface tension and other ingredients in the detergents increase the wettability of the textile. Due to the higher wetting and forced mechanical agitation, the dirt gets broken down into smaller molecules and leaves the fabric with surfactants. However, due to various reasons like reduction in surface tension, and reduction in surface friction, the role of detergent usage is noted as one of the important aspects which stimulate the microfiber shedding from the synthetic textiles.

The researcher assessed the influence of the washing parameters on the microfiber shedding behavior of the commercial fabrics. The researcher purchased fleece jackets made of 100% polyester, 100% acrylic, and 65/35% polyester cotton blend apparels and analyzed the impact of the laundry. They evaluated the roles of detergent, temperature, and fabric conditioner, and they had kept the washing duration and spinning speed constant for all the garments. The study measured the microfiber shedding with a bio detergent, non-bio detergent, and without detergent. The results revealed that irrespective of the fiber usage, the use of non-bio detergent sheds more number of microfibers than the fabrics laundered with a bio detergent and without detergent. The significant difference is noted in the case of cotton/polyester blended fabric with non-bio detergent and the other two treatments [25]. In a similar study, to measure the fiber shedding, the researcher used fleece blankets for laundry in the presence of detergent, detergent with softener, and without any additives. The microfiber shedding was noted high in the initial wash and subsequently reduced till the fifteenth wash. Hence, researchers used the average of fiber shedding value from wash number 8, 9, and 10 followed by drying, to represent the long-term emission. The results reported that a weight-loss percentage of 0.00108 wt% noted for the laundry with no additives, 0.00140 wt% for the laundry with detergent, and 0.00124 wt% for detergent + softener treatment. Based on the results, the researchers noted a very minimal effect of detergent usage on the microfiber shedding. These results were in contradiction with the previous researcher [25]. However, the study further enlightened the effect of the drying process. The experiment results indicated that compared to the laundering process, the spin-drying process imparts lots of physical stress on the materials and this makes the fabric shed 3.5 times more microfibers than the washing process [26]. The previous textile literature reported that the addition of detergent/surfactant into the laundry process significantly reduces the mechanical action of the laundry process. The development of foam, due to the use of detergent along with the absorption of the surfactant on the surface of the fiber reduces the mechanical stress applied to the fabric. Thus, the use of detergent reduces the inter-fiber friction inside the structure and restricts the fiber damage [27]. This might be the possible other reasons for the reduced microfiber shedding in the presence of the detergent.

Linn Astrom assessed the impact of the detergent in microfiber shedding as a part of his research dissertation. Except for the fleece fabric, all other fabrics were developed and dyed in the laboratory itself to have control over the properties. He used liquid detergent to quantify the effect on fiber shedding against different fabrics of polyester, polyamide, and polyacrylic type, including fleece and microfleece. The results strongly indicated that the use of detergent significantly increases the shedding of the microfiber from the fabric [28]. In a comparable work, interlock and single jersey 100% polyester fabrics were washed using liquid detergent, powdered detergent, and de-ionized water. Though the researcher did not notice any significant effect of fabric types on shedding, the detergent usage showed a significant increment in the microfiber shedding than the de-ionized water. Irrespective of the type of detergent used, the shedding rate was noted very similarly in both cases. When the researcher evaluated the effect of washing temperature and laundry time, they did not find any effect of these parameters on microfiber shedding. Similarly overdosing of detergent quantity also had no influence but at the same time, the increment in the surfactant concentration in the wash liquor had a major impact on shedding than other parameters [29]. Their study also confirmed that the mechanical stress generated by the laundry process is not one of the main reasons for shedding. This is because increment in laundry time did not yield a higher amount of microfiber shedding. The increment in the mechanical stress for a longer washing time produced the same amount of shedding similar to lower laundry time. Hence, the researchers concluded that the use of surfactants (detergents) could be the main reason for fiber shedding, as it mobilizes the broken fibers from the fabric surface to the liquid phase [30].

Almroth et al. determined the effect of detergent usage in microfiber shedding behavior with different polyester fabrics, fleece, and microfleece fabrics. The results showed a significant increment in shedding with the use of commercial detergent compared to the washing without detergent. The increment in fiber shedding with the addition of detergent is mainly associated with the reduction of surface tension in the presence of the detergent. The basic nature of detergent or surfactant present in the detergent is to increase the wetting by lowering the surface tension. The reduction in surface tension increases the wettability and so releases the trapped microfibers from the structural spaces. In addition to that, the secondary function of the detergent is to disperse the dirt and dissolve the dirt and stains in the water. This property of the detergent helps in the improved transportation of shed microfibers from the surface of the textiles and so the results showed a higher count of microfiber on laundry with detergent [13]. In a study with polyester (woven and knitted fabric) and polypropylene fiber, researchers measured the impact of detergent on microfiber shedding. The results showed an average of 162 ± 52 microfibers per g of fabric in washing, it increased to 1273 ± 177 for liquid detergent, and for powder detergent, shedding is noted as high as 3538 ± 664. The researcher suggested that the higher release of microfiber in the presence of powdered detergent might be attributed to the presence of water-insoluble inorganic contents in it. Zeolite is a perfect example, which is commonly used in detergents. These compounds will abrade on the surface of the fabric and creates friction. The powder detergents are generally considered to be stronger in removing soils and stains from textiles due to this alkali-based nature. Hence, during the laundry, the addition of powdered detergent changes the pH drastically, and this damages the polyester fiber by slow alkalization [27]. It also mentioned the possibilities of potential errors in the case of powder detergent due to the presence of insoluble detergent in the filters [29].

The researcher estimated that based on the lab-scale reading, a 5 kg typical wash load of polyester can able to shed up to 6,000,000 to 17,700,000 (0.43–1.27 g) based on the type of detergents used in the laundry. The results of the research are provided in Fig. 4 [29]. Cesa et al. used liquid detergent and evaluated the microfiber shedding from cotton, polyester, polyamide, and acrylic fabrics. In contrast to the previous research works, their results revealed that the use of detergent reduced the microfiber shedding significantly in the case of synthetic fiber. Though cotton fiber showed a reduction in shedding, the difference between with and without detergent

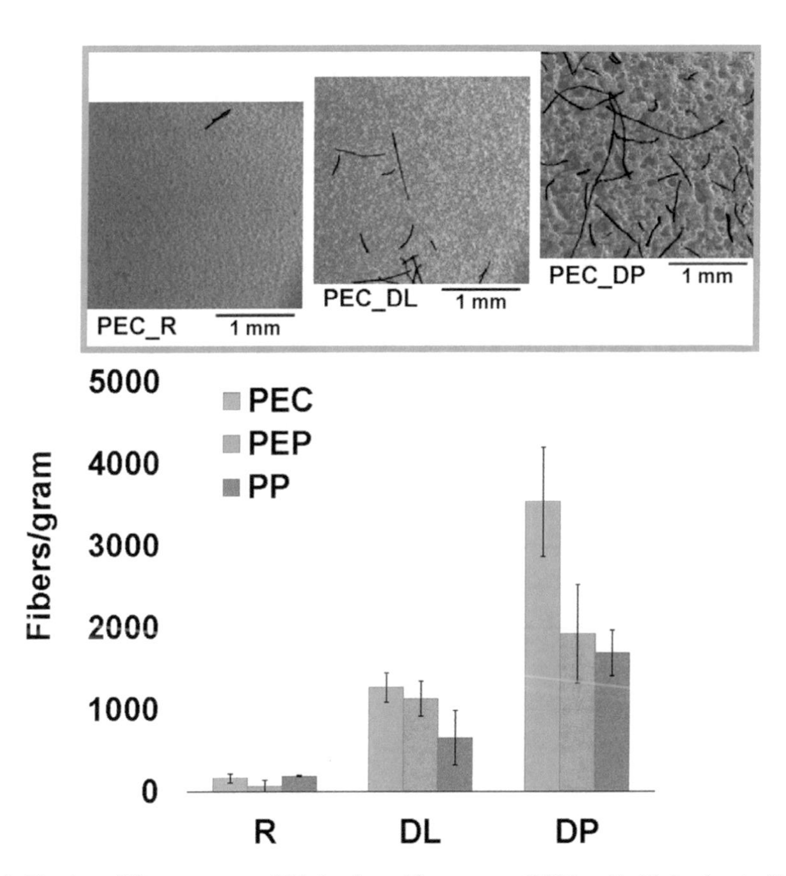

Fig. 4 Number of fibers per gram of fabric released from woven (PEC) and knitted polyester (PEP) and woven polypropylene fabrics (PP), during domestic washing simulations performed with water (R), liquid detergent (DL), and powder detergent (DP). In the upper part of the figure, SEM images of the filters were collected by simulating washings of PEC with water, liquid detergent, and powder detergent as reported by De Falco [29] (Reprinted with permission)

is negligible. The researcher came up with a different hypothesis that describes the hydrophobic nature of the synthetic textile over cotton fabric. Due to the hydrophobic nature, the synthetic textile did not get wet and the presence of surfactant from detergent reduced the friction with fabric due to the lower surface tension. This might be the reason for lower microfiber shedding in the synthetic textile in the presence of detergent compared to the washing process without detergent. The researcher also pointed out that, they have conducted testing in commercial laundry machines over the laboratory testing machine. Unlike the colourfastness and tensile test kind of tests, the laundry process in a laboratory did not represent the real-life situation as the commercial washing machine does. The laboratory machines act more harshly on the fabric. The researchers pointed out that this might be the reason for the difference in microfiber shedding in the presence of the detergent [31].

While evaluating the laundry parameters, Libiao Yang et al. measured the effect of liquid detergent applications on microfiber shedding. They have used polyester, polyamide, and acetate fabric on the laundry process and their results reported a significant increase in the microfiber shedding with detergent usage. Compared to the polyester, amide and acetate fabric shed more amount of microfibers in the laundry as represented in Fig. 5. The research justified that the deposition of surfactants on the fiber surface reduces friction and increases the shedding. They also commented on the fact that the addition of detergent also increases the pH of the solution that also had a significant influence on the fiber properties [32]. This is in contrary to the other researcher who reported reduced friction reduces the fiber damage [27]. It is also necessary to remember the previous researcher's findings that neither the composition of the detergent nor the quantity has no impact on the microfiber shedding [13]. In support of existing results, other researchers used standard liquid detergent for

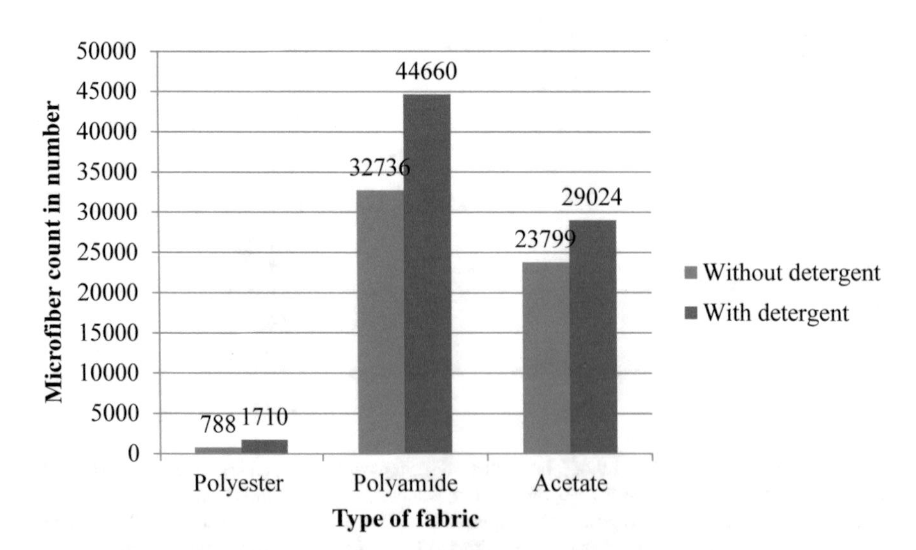

Fig. 5 Effect of liquid detergent usage on microfiber shedding quantity (Reprinted with permission) [32]

the laundry process. Upon the laundering process, the selected fibers like cotton, polyester, rayon, and polyester-cotton fabric showed an increased fiber shedding with respect to the detergent usage. The results were noted statistically significant for all the fabrics except for the rayon fabric [33].

While evaluating the microfiber shedding of fleece garments with and without detergent, researchers used a detergent pod, as per the European regulation, and followed standard conditions. The results showed no significant difference in microfiber shedding between the laundry with and without detergent. Though the results were in line with previous researchers [26, 30], the researchers mentioned that the difference in shedding might be due to the type of laundry process used. The usage of a commercial laundry machine over the lab-scale laundry machine might be the most important reason for insignificant results [34]. A report by Kelly et al. mentioned that the addition of detergent did not change the amount of microfiber shedding. Interestingly, the researchers also used commercial liquid detergent in their study. However, they did not report anything about the detergent type but they reported that their deviated results from previous research [29, 30, 33] might be due to the use of normal laundry. They claimed that the use of steel balls-related washing method might be the reason for higher microfiber shedding in the mentioned research. The use of steel balls might have increased the interaction between the textile structure and surfactant and elevated the microfiber shedding [35]. Recent research from Columbia University's Lamont-Doherty Earth Observatory reported that laundering is one of the important sources of microfiber pollution. They have reported that the addition of detergent in the laundry increases the microfiber shedding up to 86% than the laundry with pure water. It is mainly associated with the mechanism of the detergent, which interacts with fiber and yarn and loosens the structure for better cleaning. This enhances the shedding of microfiber from the internal structure of the textile. Further to add they have reported that polyamide fabrics shed more microfiber out of all the 32 types they tested [36].

To summarize the effect, overall most of the research results showed that use of the detergent significantly increased the microfiber shedding [25, 28, 32]. However, the difference in the shedding is noted between various types of detergents namely, non-bio detergents and powder detergent showed a higher microfiber shedding than the liquid detergent or bio-based detergents [25, 29]. The studies which showed a lesser or insignificant difference in microfiber shedding mostly used liquid detergents in their analysis [27, 35, 37]. All research works confirm the role of surfactant in the microfiber release both in the negative (more shedding) and positive aspects (less microfiber shedding) [27, 30]. Similarly, the concentration of the detergent used in the laundry did not show the effect on microfiber shedding [13]. It can be noted that the different detergents behave in different ways based on the formulations used in the detergents and few researchers also reported the impact of the washing methodology followed [31, 34, 35].

2.2 Effect of Laundry Temperature on Microfiber Shedding

Laundry temperature is often noted as an important parameter in the laundry as it sometimes increases the chemical reaction between the laundry ingredients and the textile structure. Hence, few researchers also evaluated the effect of laundry temperature on the microfiber shedding of synthetic textile during the laundry. The first research performed by Napper and Thompson estimated two different washing temperatures namely 30 and 40 °C. Though acrylic fabric sheds more microfibers at 30 °C, it is noted that the polyester fabric sheds more fiber than acrylic at 40 than 30 °C. The results were statistically insignificant but the researchers reported the parameter is of vital importance [25]. While measuring the shedding behavior of knitted and woven polyester fabric and polypropylene, researchers used different temperatures, respectively, 40 and 75 °C. Though they did not perform any detailed analysis of different fabrics, the researchers commented that increment in washing temperature increased the microfiber shedding. The statistical analysis performed by the researcher did not show any significant difference in the temperature impact. However, it was explained by researchers that the higher temperature in washing might have increased the surface hydrolysis of the synthetic textile due to the higher alkaline nature of the detergent [29]. In a different study, researchers measured the microfiber shedding at two different temperatures against cotton, polyester, rayon, and cotton/polyester blend fabric. The researchers selected two different laundry cycles from the washing machine namely cold (25 °C) and warm cycles (44 °C). On the analysis, the results reported that the cotton and cotton blend material showed a significant increment in the fiber shedding with the temperature. While comparing it with cotton, the synthetic fiber (polyester) did not show any significant increment in shedding with the temperature changes. The changes were significant when the temperature changed in the presence of the laundry detergent [33].

In a detailed analysis, research results measured the temperature effect on microfiber shedding in the presence of detergent and absence. Researchers assessed the shedding behavior of polyester, amide, and acetate fiber at 30, 40, and 60 °C without detergent. The findings of the study revealed that the increment in temperature increased the release of microfiber from the synthetic cloth. The results showed that in the case of polyester fabric, there is a slight increment in the shedding noted between 30 and 40 °C (from 788 to 1032). However, the microfiber shedding significantly increased when the temperature increased to 60 °C. At 60 °C temperature, the polyester sheds 13960 ± 2406 fibers but whereas at 40 °C the count was noted as 1032 ± 150. Though polyamide and acetate fabric sheds more amounts of microfibers than polyester, they showed a very little increment in the count noted comparatively less than polyester fabric. The findings of the results were in line with the previous research [33] who reported a similar result. The author reported that the increment in temperature at a lower level (up to 40 °C) did not have any impact on the microfiber shedding. Hence, the researchers suggested a low-temperature laundry for the synthetic textiles washing, to reduce the microfiber shedding [32]. The details of the microfiber count at different temperatures at plain water are provided in Fig. 6.

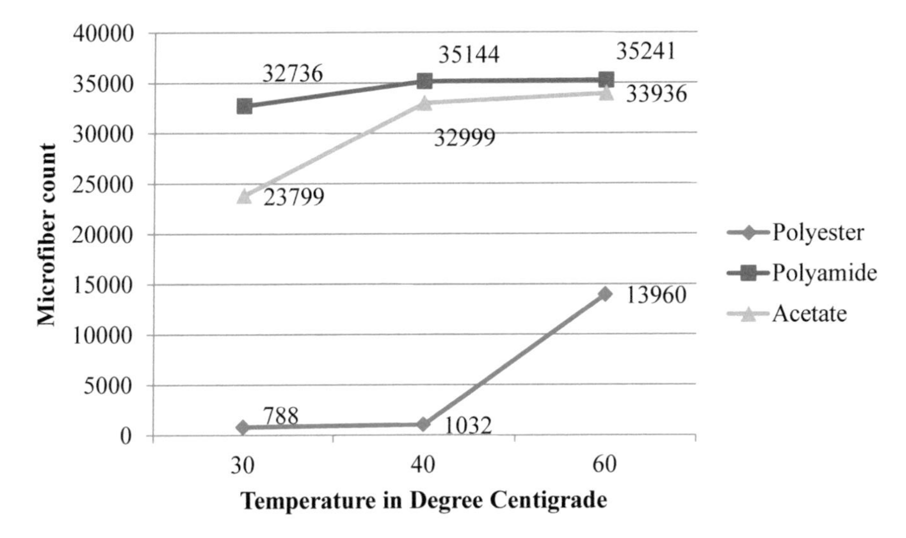

Fig. 6 Amount of microfiber identified in the wash liquid at different temperatures [32] (Reprinted with permission)

Lant et al. evaluated the effect of cold express cycles in the European washing machine on the microfiber shedding by comparing it with 40 °C cycle as per the European washing procedure. The researchers used a similar cloth load for both the cycles and the same liquid detergent for both processes. The results showed that the use of the cold cycle reduced the microfiber shedding up to 30% reduction with a statistical significance of 95% confidence level. The 40 °C cycle showed an average release of 181.6 ± 87.1 ppm but in the case of the cold cycle, it is noted as 129.5 ± 42.9 ppm. The details of the findings are provided in Fig. 7. The researchers concluded that the use of lower temperature washing cycles significantly reduces the microfiber shedding and recommended the benefits of it [34]. In very similar research, Cotton et al. reported that the cold wash procedure (in a mixture of both synthetic and natural fiber) showed relatively lower microfiber shedding than the 40 °C temperature wash. However, the author did not perform any separate investigations on the shedding behavior of the synthetic textiles [37].

Other researchers measured the microfiber shedding of 100% polyester with lower temperatures of 15 and 30 °C with an average laundry time of 15–60 min. The results reported that either the reduction in the temperature or laundry duration did not show any impact of the microfiber shedding. The researchers selected lower temperature (15–30 °C) as most of the previous studies represented higher temperature ranging from 40 to 80 °C. The results showed no significant difference in microfiber shedding at a cold temperature below 30 °C [35]. These findings were in line with the previous research result reports [34, 36]. When the polyester is treated with different temperatures at 25, 40, 60, 80 °C, the research did not find any increase in the microfiber shedding. However, the addition of liquid detergent slightly increased the microfiber shedding at higher temperatures (60 and 80 °C). The combined influence of the water

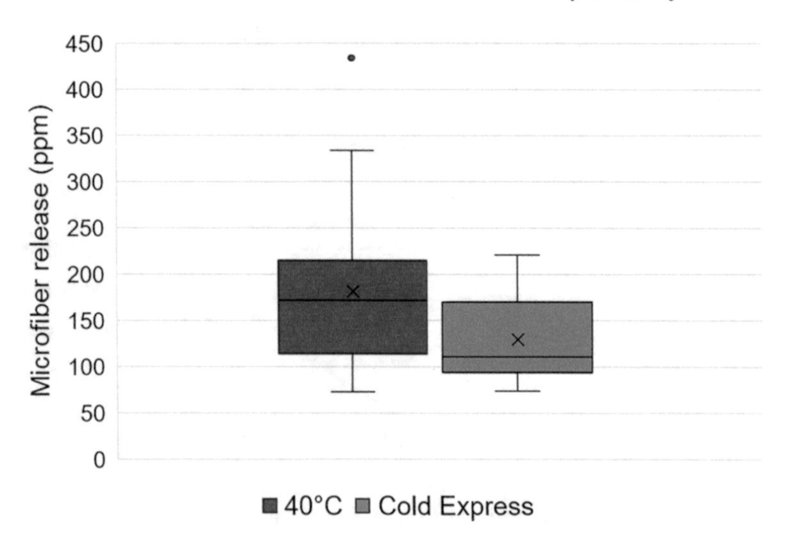

Fig. 7 Boxplot of microfiber release from soiled consumer wash loads in a 40 °C cycle ($n = 19$) with cold express cycle ($n = 19$) (Reprinted with permission) [34]

temperature and water in the laundry directly has more impact on the glass transition temperature of the fiber. The higher crystalline nature and hydrophobic behavior of the polyester prevents it from swelling but this nature in front of higher temperature breaks the individual fibers often. The authors justified that the lower shedding at lower temperatures might be due to the lesser effect of the alkaline pH of the solution on synthetic textile, specifically polyester [30]. The other synthetic textiles like polyamides are mostly affected by oxidative attack [26].

The results of the research confirmed that generally, the increment in the laundry temperature had a significant effect on microfiber shedding from the synthetic textiles and more specifically polyester. Several research works performed below room temperature (<30 °C) showed a significant reduction in shedding. Irrespective of the temperature used, all the research work generally reported an increased tendency in microfiber shedding in the presence of the detergent. Based on the findings, several researchers suggested laundering synthetic clothing at a lower temperature significantly reduces the microfiber shedding.

2.3 *Effect of Laundry Cycle/Time on Microfiber Shedding*

Along with detergent and temperature, the most commonly addressed parameter which influences the microfiber shedding is the laundry duration. As the microfiber shedding majorly attributed to the mechanical stress imparted by washing it is directly related to the laundry time. Hence, several researchers have investigated the effect of laundry time or laundry cycle on the microfiber shedding. The general trend noted in all the research showed a reduction in microfiber shedding with increment in the number of cycles. The first study on the effect of washing cycle or time on the microfiber shedding was performed by Napper and Thompson. They examined the acrylics, polyester, and polyester/cotton fabric for 4 consecutive laundries. The researcher noted a decrease in the microfiber in all the cases. In the case of polyester, the first wash shed 2.79 mg of polyester and at 4th wash, it was noted as 1.63 mg. Compared to the polyester, for acrylic fabric, shedding reduced more but followed a similar pattern. The acrylic fiber count has been noted as 2.63 mg initially and at the 4th wash, it was 0.99 mg [25]. Similarly, a study explored different polyester fabrics (anti-pilling fleece, non-anti-pilling fleece, edge covered fabric from softshell and tech sports t-shirts) and cotton fabrics (jeans and shirts) for their fiber shedding manners concerning the wash cycle. The shedding was noted high in the first laundry up to 0.12–33% weight per fabric weight. The second wash results showed a significant level of reduction in the polyester fabric shedding approximately one-third of the mass got reduced. The shedding level remained the same in the third to fifth wash for anti-pilling fleece and non-anti-pilling fabric. However, there was a one-tenth loss in the case of tech sport and softshell fabric in the fifth wash. The average number of microfibers released ranges from 210,000 to 13,000,000 per kg of the fabric in the wash. Out of all the fabrics used, the higher release was noted with softshell and tech sports. There was a reduction of one-tenth of microfiber in the first 5 washes for all the fibers used except for non-anti-pilling fleece. Figure 8a represents the microfiber images as reported by the authors and Fig. 8b represents the microfiber shedding number per wash in the case of polyester fabric [38].

In order to question the real-time effect of microfiber shedding on knitted polyester clothing, Almroth et al. studied shedding behavior at different laundry cycles. They selected two knitted polyester fabrics randomly as new and older material. The repeated washing process was performed on both fabric (1, 2, 5, and 10 washes). The

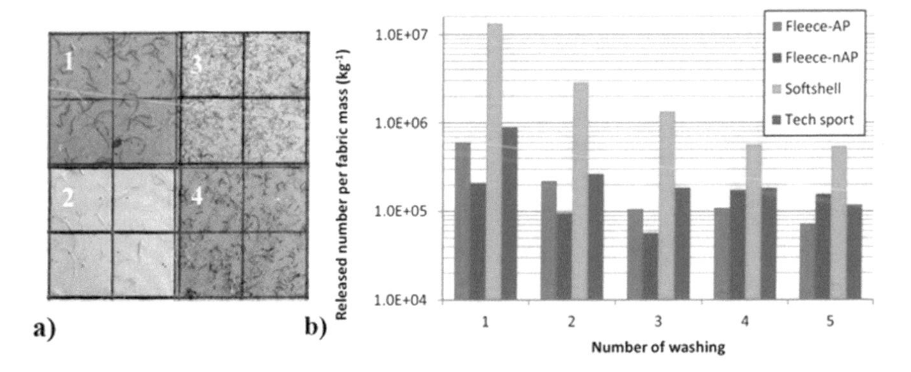

Fig. 8 **a** Images of fibers filtrated from the first washes. 1. Anti-pilling fleece. 2. Non Anti-pilling fleece. 3. Edge covered fabric from Softshell. 4. Edge covered fabric from techsport t-shirts. **b** The numbers of microfibers released from polyester fabric per fabric mass (per kg) in five sequential washes (Reprinted with permission) [42]

results indicated that the new garment sheds more number of fibers than the older garment. The older garment did not show a reduction in microfiber shedding after 2, 5, and 10 washing cycles. Whereas, in the new garments, shedding significantly reduced in the subsequent washes [13]. Similar research performed three laundry cycles to evaluate the microfiber shedding of polyester, rayon, a polyester-cotton blend, and cotton fabric. The results showed a significant reduction in the microfiber shedding than the preceding wash until the third wash. However, even after the third wash, the fabric shed a reasonable amount of microfibers [33]. Researchers also calculated the microfiber detachment rate using different polyester fabric. They evaluated the shedding nature in the preliminary trial and found that after the 5th wash the shedding was stabilized. Hence the researcher measured the microfiber shedding at the 5th wash as the long-term release quantity. The researcher evaluated the shedding of polyester fluffy fabric, polyester normal fabric, polyester/elastane, and polyacrylic fabric. Out of all the selected materials, higher shedding of 465,000 MF/m^2 was noted with acrylic knitted fabric followed by polyester fluffy fabric (30,000 MF/m^2). The researcher noted a higher release of the microfiber (MF) in the first washing than any other wash [39] (Fig. 9).

For cotton, polyester, polyamide, and acrylic fabrics, the shedding actions were investigated for 10 consecutive cycles with and without detergent addition in the laundry. By supporting the previous researchers' findings, the researchers reported on 10 cycles, the microfiber shedding stabilized after 5 washes. From the sixth wash onward, irrespective of the fabric type the microfiber shedding stabilized in the case of synthetic textiles. Out of the selected synthetics, acrylic fabric shed more microfibers (10 washes) and the first three washes contributed 37–76% of the mass released during the 10 washes. Polyacrylic fiber released the lowest value of 37–41% in the first three wash and polyamide accounted for the maximum value [31]. In a different study, Francesca De Falco et al. measured the microfiber shedding of two different garments sourced from commercial outlets. One garment made of 100% polyester and 2nd garment made of polyester (front panel), and cotton/modal

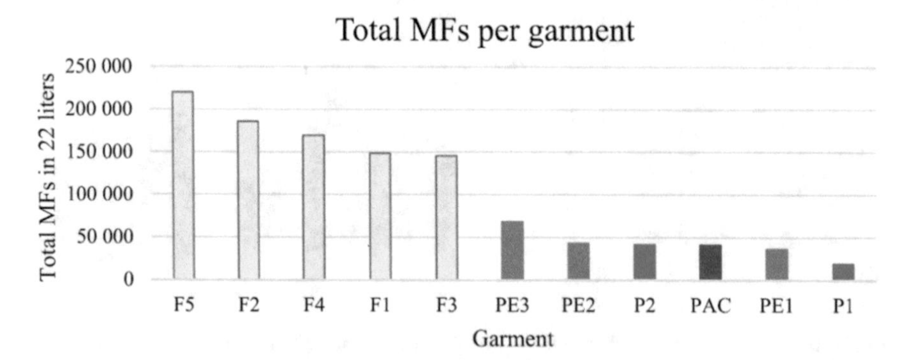

Fig. 9 Represents the total microfibers in 22 L of wash liquid (F1–F4, fluffy woven fabric, F5 knitted fabric, PE—polyester elastane garment, PAC—Acrylic polyamide and P—knitted (1) and woven (2) polyester garments) [39] (Reprinted with permission)

(back). The 10 consecutive washing resulted in a decrease in the fiber shedding in all the cases. However, for the 1st garment, the researcher noted a decrement in the microfiber shedding till the 5th wash. After the 5th wash, the 100% polyester shedding almost became stable but the 2nd garment showed a higher shedding. The researcher supported that this might be due to the presence of cotton and modal content [24]. Claire O'Loughlin measured the microfiber release quantity of polyester with elastane sportswear and nylon swimwear. The researcher evaluated the shedding for 15 laundry cycles. They reported 41% of the shedding for both the textiles. Irrespective of the garment or detergent usage, after the 5th wash no difference was noted in the release quantity. However, the results showed a reverse trend in the 11–15th washes where the higher release percentage (31%) than the non-detergent added set was observed. These findings again suggest having a different look at the abovementioned results [40]. A recent study evaluated the effect of laundry cycles up to 10 washes using 12 different textile samples. The results revealed that 6–120 times higher microfiber shedding was noted in the first wash compared to the 10th wash. A maximum fiber release was noted in the first three washes and a further reduction in shedding was noted in the further washes. The release count stabilizes in the 5th wash with 10–1200 fibers/g of textile washed. Concerning the fiber length released, the first cycle released shorter fiber than the 10th cycle [41].

To recapitulate the results, it is noted that most of the research results mentioned that the microfiber shedding quantity reduces in the consecutive laundry cycles [23, 30]. Many researchers accepted that the major release of the microfiber occurs in the first wash in the case of synthetic fiber [30, 42]. The microfiber release quantity stabilizes after the 5th wash in the case of all synthetic fiber irrespective of the detergent usage and fabric structural parameter [24]. Though few researchers reported an insignificant impact of detergent, the majority of the researchers reported the impact of detergent as a significant parameter [40].

2.4 *Effect of Softener Treatment on Microfiber Shedding*

Fabric softeners are commonly used in the laundry to soften the fabric after the laundry process. During the laundry process, the washing machine abrades the materials mechanically and forms fuzz by opening the entangled fibers. The produced lint fibers on the surface become stiff due to the drying. Further to add, the use of detergent removes the lubricants like waxes and oils from the fiber material, and the material becomes hard and scratchy to wear. Hence to impart lubrication between fibers and to make the fabric softer after laundry the softeners are generally used [41]. Besides, in the case of synthetic fabric, softeners are used to avoid static charge build-up on the fabric. However, regarding the effect of softener treatment on the microfiber releasing quantity, there are no clear studies available to date. Most of the studies performed to evaluate the performance of the different fiber types, detergent, and temperature. Often softeners are considered as optional material during the laundry, hence no emphasis is given on the effect of softener treatment in washing on the microfiber shedding pattern.

With respect to the microfiber shedding, Francesca De Falco et al. investigated the impact of fabric softeners used in the laundry. The results of the research revealed that, when compared to a normal laundry with distilled water, laundry with softener significantly reduced the amount of microfiber releasing quantity. Further, the author reported that since the pH of the washing liquor was maintained common for both the cases (with softener and without softener), the influence of the pH is not significant on microfiber release. The findings showed that the use of softener with liquid detergent in a household washing of polyester fabric reduced the number of microfiber shedding to 4,000,000 microfibers (approximately), compared to a wash with the same condition without the softener (6,000,000 microfibers). The researcher reported a 35% reduction in the microfiber release from the polyester fabric due to the use of fabric softener in the laundry. The researcher justified that the addition of softener reduced the friction between fibers and made to arrange parallel in the fiber bundle and also by providing softness, reduces the damaging of fibers [29]. To support this finding, older textile research details that the addition of softener in the rinse cycle reducing the pilling of polyester fabric. The researchers used polyester knitted and woven fabric with softener in the rinse cycle. After ten cycles of wash, the results showed a reduced pilling on the surface of the polyester fabric (with a higher rating of 5) washed with the softeners. Similarly, the researcher also noted a fabric breaking strength reduction of 13–17% after the softener treatment. The lubrication effect and reduced friction developed by the softener reduced the breaking strength and so prevented the fiber breakage and so reduces the pill formation in the polyester fabric [43, 44]. Figure 10 represent the effect of fabric softener treatment on the microfiber shedding from textiles.

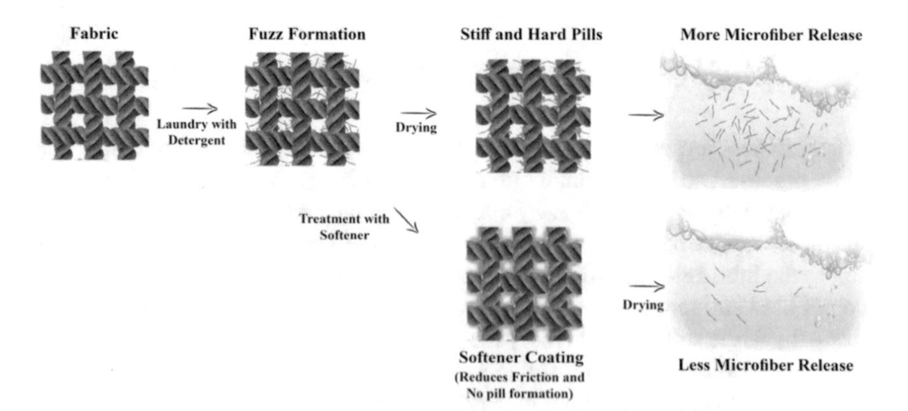

Fig. 10 Mechanism of softener action on microfiber shedding from the literature (Authors own illustration)

Few other studies were also reported insignificant results of softener treatment on the microfiber shedding. Though the earlier study explored the softener effect, the research did not find the impact on microfiber shedding [25]. In a study, researchers compared the microfiber shedding with, without detergent and also with detergent and softener. The results showed a microfiber shedding percentage at eighth, ninth, and tenth washes to replicate the long-term effect. The mass loss percentage was noted as 0.00108 wt% (without detergent), 0.00140 wt% (with detergent), and 0.00124 wt% (detergent + softener). The findings showed no significant difference in microfiber shedding with softener treatment. To confirm the effect, the researcher also measured the microfiber releasing quantity with 5-year-old polyester fleece. These results were in agreement with each other and confirmed the long-term effect [27]. A related negative result was also reported by Lant et al. Who studied the effect of softener treatment with different commercial fabrics (fleece and normal polyester t-shirts) to evaluate the microfiber shedding quantity. The result did not show any significant change in microfiber shedding with the use of softener. The researcher tried the laundry with European conditions and also with the North American style testing method. Both the methods showed similar insignificance of softener treatment on microfiber shedding [34].

Though very few studies examined the effect of softener on textile microfiber shedding, the results confirmed a significant effect of softener on microfiber shedding. Whereas few studies reported a positive impact on microfiber shedding (reduction) others also reported a negative impact. This suggests a research requirement in this area to evaluate the softener effect. Various researchers used different softeners and laundry methods or machines to perform the analysis. This might be the main reason for variations in the findings of those researchers. Hence, it is important to regulate laundry parameters and ingredients used in the process to evaluate the softener's effect on microfiber releasing quantity.

2.5 *Effect of Washing Machine Type/Method on Microfiber Shedding*

The type of washing machine used or the method adopted for the microfiber shedding quantity is one of the most controversial topics in this area. Several researchers used different types of washing machines in measuring the microfiber shedding quantity. Researchers claimed that the variation in their results might be due to the type of machine or method they adapted in their research. Hence it is necessary to study the importance of machine type on microfiber shedding. In a comparative study of front- and top-loading washing machines, Hartline et al. evaluated the shedding behavior of new garments and also mechanically aged garments. Their results reported that the average mass recovered from the front-load washing machine was 220 mg and the top-load washing machine was 1906 mg. The statistical analysis revealed a significantly higher mass of microfibers in the top-load machine than a front-load machine. The researcher reported that the higher agitation of the central agitation mechanism might have caused a higher abrasion on the top-load machine than the front-load. Based on the results researcher mentioned that further studies on different types of machines may be required to understand the effect of mechanical agitation. The results of the research work are reported in Fig. 11 [45].

In a comparative research of cotton, rayon, and polyester, researchers mentioned that the accelerated laundry process always yielded a higher mass of microfiber than the household laundry machine. The researcher suspected the use of steel balls might have caused higher damage to the fiber. The finding revealed approximately 40 times more mass yielded in the accelerated lab model washing machine than the commercial laundry machine. The correlation analyses revealed an 80% correlation between the microfiber yield from garments and the type of machine used. However, the researcher supported that the use of accelerated laundry is most suitable for quantitative measurement of the microfiber quantity. Due to the large size, a higher volume of water usage, and potential errors due to the microfiber deposition

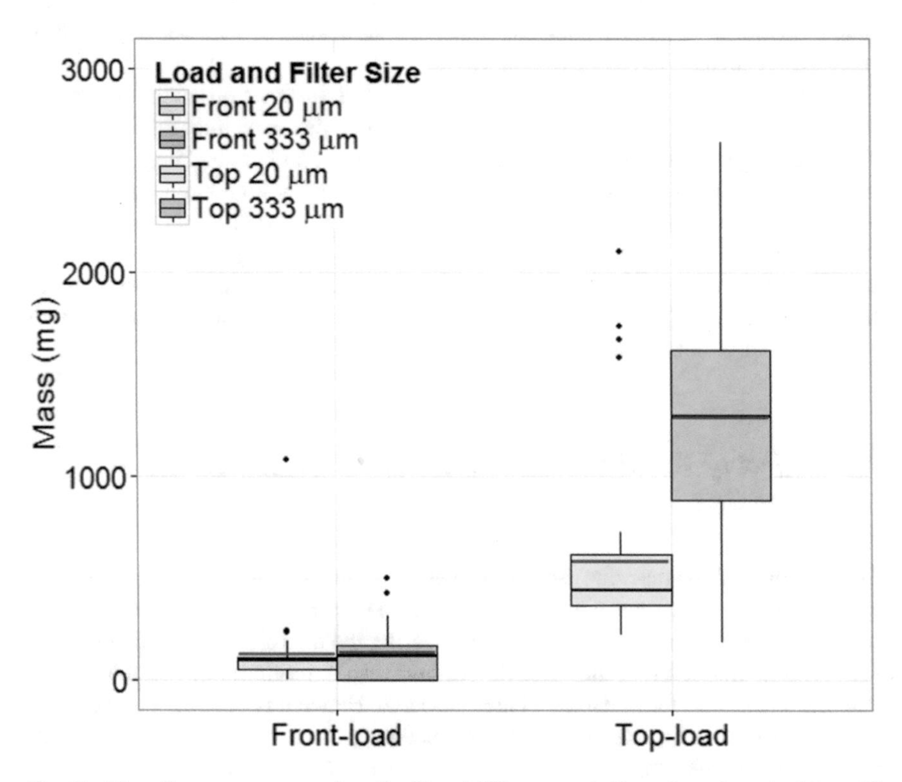

Fig. 11 Microfiber mass recovered on the 20 and 333 μm mesh filters from front-load ($n = 30$ each box) versus top-load ($n = 40$ each box) washing machine modes. Red lines indicate means, black lines medians, and black dots outliers (beyond 1.5 times the interquartile range) (Reprinted (adapted) with permission [45]. Copyright (2016) American Chemical Society)

at different components, the home washing machines could lead to poor quantitative analysis. Further, on the measurement of microfiber size, it is noted that the accelerated washing machine produced shorter microfiber length than the home washing machine. This confirms the higher agitation of accelerated washing machines that caused more breakage due to mechanical abrasion [33].

Similarly, the effect of the platen and pulsator washing machine process on microfiber shedding quantity was measured. The results revealed that a weight loss of 0.0068–0.12% was noted in polyester fabric. The loss percentages of 0.79–1.20% for polyamide and 0.58–1.01% for acetate were noted. In the case of the pulsator machine, the reduction percentage is noted higher compared to the platen machine. A weight-loss percentage of 0.0077–0.20%, 0.89–1.69%, and 0.76–1.82% were noted, respectively, for polyester, polyamide, and acetate fabric. Overall the researcher noted 1.08–2.13 times higher shedding in the case of a pulsator machine than a platen machine. These findings of the researchers are in line with the previous works, who also reported a similar effect of pulsator or laboratory machine on microfiber shedding [32]. In diverse research, the researcher used two washing machine types namely to represent the European standard (high-efficiency modern top-load) and also to mention the North American standard (top-load), a traditional top-loading washing machine. The results represented that the high-efficiency top-load machines released a comparatively lower amount of microfiber quantity than the traditional top-load washing machine. The research reported a reduction of 69.7 and 37.4% in microfiber shedding, respectively, for polyester fleece and performance t-shirt in high-efficiency washing machine compared to the traditional top-load machine. Though there are differences noted in the high-efficiency machine, the researchers believed that the higher water ratio in the traditional washing machine must be the major cause for the higher microfiber shedding. The microfiber shedding in different washing machines with detergent and detergent with softener is provided in Fig. 12 [34].

The most common dispute among the research community is the use of laboratory-scale washing machines like laundrometer, Gyrowash, or Washtec P. The laboratory-scale machine uses steel balls for laundry hence many researchers believe that this harsh treatment creates higher shedding. Similarly, whenever, the front-load and top-load washing machines are considered, the top-load machines always released more amount of microfibers than the front-load. The other study represented that the use of Gyrowash protocol provided regular household laundry results [30] and Kelly et al. reported that a tergotometer provided a better understanding of small-scale simulations [35]. The following Table 1 consolidates the existing studies and microfiber release quantity with different washing machines.

From Table 1, it can be noticed that the majority of the studies did not compare the effect of the type of washing used for the microfiber release estimation. No similar research results obtained from the existing studies as each study used a different standard and method to simulate the real-time washing. Two studies confirmed that the use of a top-load machine sheds more microfibers than the front-load machines [32, 43]. Out of top-load, research confirmed that the new machine performed better

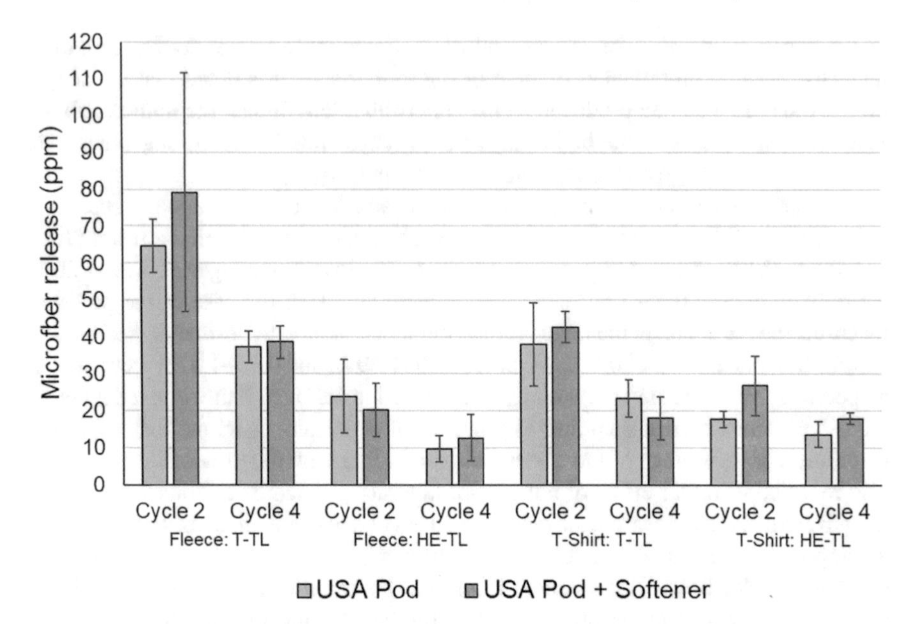

Fig. 12 Impact of fabric softener on microfiber release from polyester fleece and performance t-shirt ($n = 64$). Traditional Top-Loader (T-TL) and High-Efficiency Top-Loader (HE-TL) washing machine [34]

than the traditional machine and released fewer microfibers [34]. Few reported higher microfiber shedding with a laboratory model against standard conditions and others reported the same in household laundry machines [24, 25, 28, 29]. The data listed here showed a need for standard procedures for the study, as every researcher used different variables in their washing process. However, a high microfiber shedding is noted in the laboratory model compared to the household models based on the working nature and types of standards followed in the analysis. Hence it is important to standardize the machine setting across the world to a proper estimation of microfiber shedding.

2.6 Other Influencing Parameters Influencing Microfiber Shedding

Other than the abovementioned research works, few other researchers estimated the effect of a few other parameters like water hardness and liquid to fabric ratio in the laundry on the microfiber shedding process. This particular section details those parameters for a better understanding of microfiber shedding.

Table 1 Research works on different types of washing machine and microfiber shedding

S. no	Type of washing machine	Impact of washing machine	Quantity or percentage of microfiber shedding/wash	References
1.	Gyrowash	Not discussed	210–72,000 Microfiber/g textile per wash	Cai et al. [41]
2.	Traditional top-load and High-efficiency top-load	High-efficiency top-load sheds less	Fleece 69.7% and t-shirt 37.4% reduction in a high-efficiency machine	Lant et al. [34]
3.	Top-load, mini-washing machine	Not discussed	Minimum of 49.8 mg to a maximum of 307.8 mg of release	Cesa et al. [31]
4.	Platen and pulsator laundry machine	Pulsator machined showed higher shedding	Increased shedding in (17.6%) in pulsator machine than the platen	Yang et al. [32]
5.	Top-loading, commercial, heavy-duty washer	Not discussed	0.1–1 mg/g fabric)	Zambrano et al. [33]
6.	Bosch washing machine	Not discussed	124–308 mg for kg of fabric (from 640,000 to 1,500,000)	De Falco et al. [24]
7.	Tergotometer, and front-loading washing machines	Delegate washing sheds more fiber	800,000 microfibers in the first wash	Kelly et al. [35]
8.	Linitest apparatus	Shedding increases with the number of steel balls	6,000,000 fibers per 5 kg of polyester load	Hernandez et al. [30]
9.	Gyrowash	Not discussed	Fleece −7360 fibersm^{-2}L^{-1} and polyester fabrics which shed 87 fibersm^{-2}L^{-1}	Almroth et al. [13]
10.	Washtex-P Roaches laboratory washing machine	Not discussed	0.1 mg microfiber/g of textile	Hernandez et al. [30]
11.	Gyrowash	Not discussed	Fleece and microfleece released more fibers	Astrom [28]
12.	Front-load washing machine	Not discussed	210,000 fibers/wash	Sillanpää and Sainio [42]

(continued)

Table 1 (continued)

S. no	Type of washing machine	Impact of washing machine	Quantity or percentage of microfiber shedding/wash	References
13.	Front-loading washing machine	Not discussed	0.0012 wt%. weight loss in the garment	Bishop [27]
14.	Front-loading machine	Not discussed	700,000 fibers per 6 kg	Napper and Thompson [25]
15.	Top-loading and front-loading commercial, heavy-duty washer	Top-loading shed more than a front-loading machine	0.3% of garment mass (up to 2 g)	Putchta [43]
16.	Commercial front-load	Not discussed	>1900 fibers per wash	Brown et al. [5]

2.6.1 Water Hardness

There are no detailed studies performed on the effect of water hardness on the microfiber shedding. However, Francesca De Falco et al. evaluated the effect of hard water and distilled water for the laundry process with detergent and other ingredients along with a few other parameters like different detergents. The results of the study showed that changes in water hardness or the addition of hard water in the laundry increased the microfiber shedding compared to the distilled water cycle. Based on the previous literature on cotton, the researcher hypothesized that the hard water might have created higher abrasion on the fiber and damaged the fiber surface. This could be the reason for higher microfiber shedding in the presence of hard water [29].

2.6.2 Water Quantity or Volume Used in Laundry

Like the water hardness, the recent research work pointed out the role of washing water volume on the microfiber shedding. On analysis, the researcher used a tergotometer to simulate the front-load laundry machine and performed a small-scale analysis. In their study, they performed washing with different water volume and mechanical agitations namely 300 mL/200 rpm, 600 mL/200 rpm, 300 mL/100 rpm, 600 mL/100 rpm with 30 °C temperature and for 60 min with detergent. The results showed a significant difference in microfiber shedding. Out of all the samples, higher water volume with lower agitation of 100 rpm showed a maximum amount of microfibers compared to other styles. The research reported that more than mechanical agitation, higher water volume causes more shedding. The researcher also evaluated the same with real-time wash cycles. In a front-load washing machine the researcher used four different washes namely, I type—30 °C/cotton short/without detergent, II type—30 °C/cotton short/detergent, III type—cold express/detergent, and IV type—30 °C/delicate/detergent. The results are provided in Fig. 13. The results from large-scale investigation showed a similar trend as tergotometer analysis and showed that the delegate cycle in the front-load washing machine can shed more microfiber due to higher water volume. The result showed that approximately 114 mg or 800000 microfibers were produced in the first wash [35]. While other researchers report the mechanical stress created by laundry causes higher microfiber shedding [32, 43], these findings were noted as important.

The researchers mentioned that higher water volume causes higher hydrodynamic pressure on the fabric structure. Due to the higher surface area of the microfibers, it undergoes a higher amount of viscous forces during the laundry process. As the laundry time was significantly longer, these forces act on individual fibers and pluck the fibers from the structure. Further, the delicate cycles shed more fiber in consecutive washes, it can be understood from the above-said mechanism, the forces will

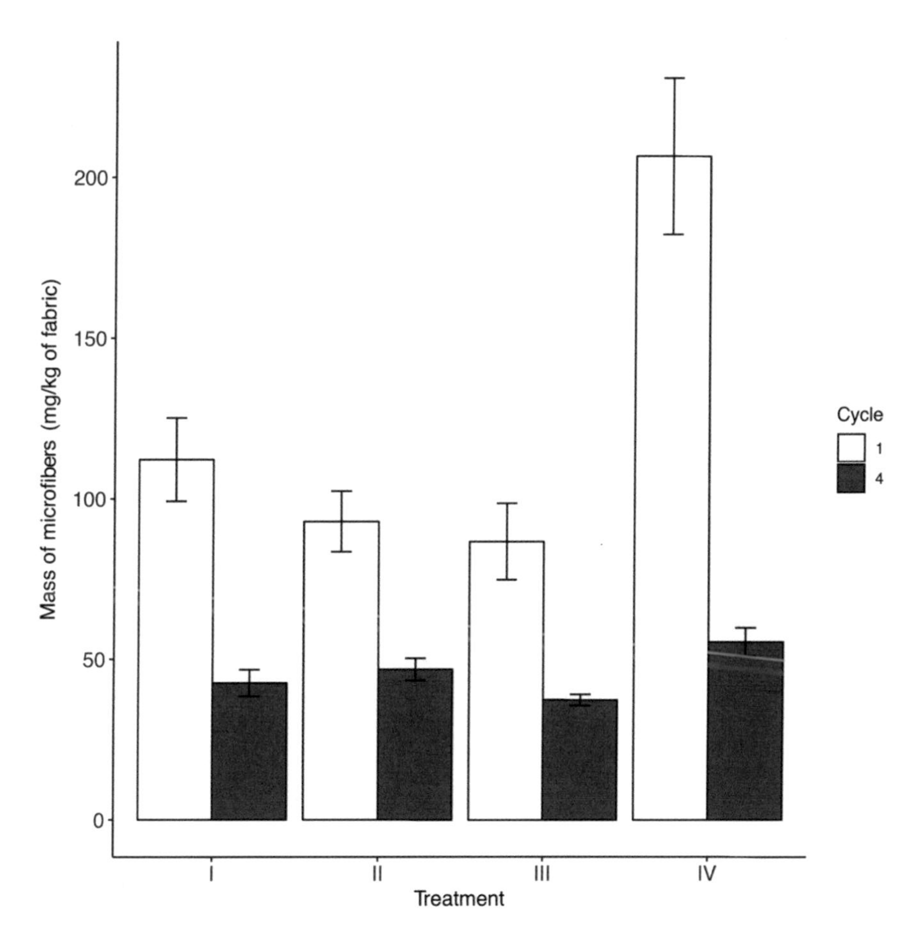

Fig. 13 Mean mass of released microfibers across four different treatment types (Reprinted (adapted) with permission [35]. Copyright (2019) American Chemical Society)

weaken the fabric structure in a period [35]. These findings also make us rethink the results of other researchers who investigated and reported that the top-load machine sheds more microfibers, as the top-load generally uses more amount water than the front-load machine [32]. This is also the same for the researcher who compared the European laundry machine and compared it with North American top-loading machines [34].

3 Summary and Recommendation

The chapter summarizes the different research work performed on the microfiber shedding analysis and details the important laundry parameters that have an impact on fiber release. From the results, it can be noted that irrespective of the fabric types all the material sheds microfiber. Due to their biodegradability, we did not bother about shedding behavior of natural fiber though it sheds more amounts. In the case of laundry use of detergent in the form of bio (enzyme oriented), power or liquid enhances microfiber shedding. The addition of softener has been mentioned to have a positive effect by reducing microfiber release. Higher temperatures and longer duration of laundry increase the shedding properties of synthetic material. In the case of washing machines, several researchers reported top-load machine creates higher mechanical abrasion and so releases more microfiber at washing. However, few research works reported the role of water hardness and liquid to fabric ratio also play a vital role in microfiber shedding. The study recommends the following points to reduce microfiber shedding in household laundry process and few points for future research.

- Use of mild detergent or bio-based liquid detergent in the laundry controls the microfiber release.
- Use of front-load washers preferred over top-load machine due to their mild mechanical action.
- Lower temperature with lower time preferred to control microfiber release from textile.
- Lower liquid volume per kilogram of fabric controls the microfiber release.
- Though softener usage was reported as positive in microfiber release, few other studies also reported opposite results. Hence further research needs to be done on different types of commercial softeners by relating microfiber shedding. Figure 14 represents the preferable laundry parameters for reduced microfiber release during laundry.

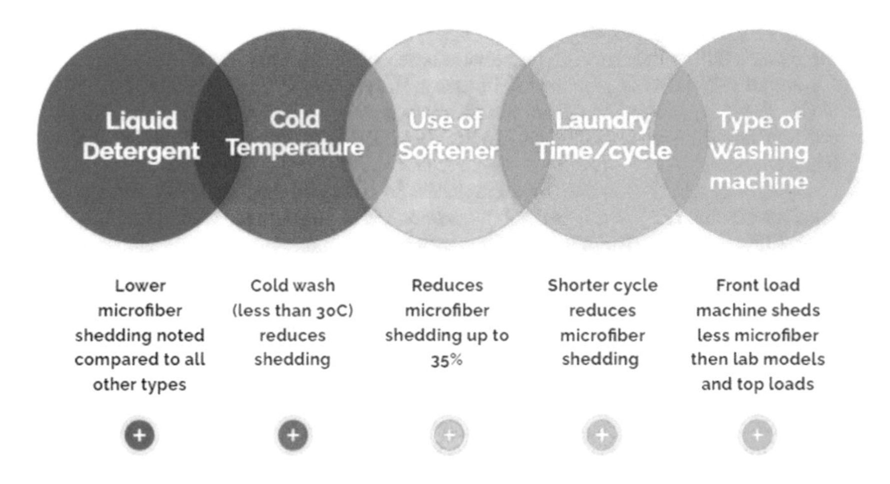

Fig. 14 Preferable parameter for environmental friendly laundry process (Authors own illustration)

References

1. Carr CM (1995) Physical and chemical effects of domestic laundering processes. In: Chemistry of textiles industry. Blackie Academic and Professional, London
2. Jerg G, Baumann J (1990) Polyester microfibers: a new generation of fabrics. Text Chem Color 22:12–14
3. Mishra S, Rath CC, Das AP (2019) Marinemicrofiber pollution: a review on present status and future challenges. Mar Pollut Bull 140:188–197. https://doi.org/10.1016/j.marpolbul.2019.01.039
4. Stanton T, Johnson M, Nathanail P, MacNaughtan W, Gomes RL (2019) Freshwater and airborne textile fibre populations are dominated by 'natural', not microplastic, fibres. Sci Total Environ 666:377–389. https://doi.org/10.1016/j.scitotenv.2019.02.278
5. Browne MA, Crump P, Niven SJ, Teuten E, Tonkin A, Galloway T, Thompson R (2011) Accumulation of microplastic on shorelines woldwide: sources and sinks. Environ Sci Technol 45:9175–9179. https://dx.doi.org/10.1021/es201811s
6. Norén F, Norén K, Magnusson K (2014) Marine microscopic debris, survey along Swedish west coast 2013 and 2014. IVL, Swedish Environmental Institute report no 2014: 52. County Administrative Board, West Götaland. https://sverigesradio.se/diverse/appdata/isidor/files/406/14638.pdf. Accessed 27 Aug 2020
7. Magnusson K, Eliasson K, Fråne A, Haikonen K, Hultén J, Olshammar M, Stadmark J, Voisin A (2017) Swedish sources and pathways for microplastics to the marine environment: a review of existing data. IVL, Swedish Environmental Institute report no 2016: C 183, Stockholm, Sweden. https://www.ivl.se/webdav/files/Rapporter/C183.pdf
8. Woodall LC, Sanchez-Vidal A, Canals M, Paterson GLJ, Coppock R, Sleight V, Calafat A, Rogers AD, Narayanaswamy BE, Thompson RC (2014) The deep sea is a major sink for microplastic debris. Royal Soc Open Sci 1(4):140317. https://doi.org/10.1098/rsos.140317
9. Ellen MacArthur Foundation (2017) A new textiles economy: redesigning fashion's future. http://www.ellenmacarthurfoundation.org/publications). Accessed 15 Apr 2020

10. O'Donnell J (2019) Are recycled polyester garments. The lowest hanging fruit for sustainability? https://thriveglobal.com/stories/recycled-polyester-garments-the-lowest-hanging-fruit-for-plastic-sustainability/
11. Mahesh P, Mukherjee M. Microplastics the most widespread and persistent hazards of plastic. Toxics Link Factsheet Number 56
12. Microfibers (2019) Beachapedia. http://www.beachapedia.org/Microfibers
13. Almroth BMC, Åström L, Roslund S, Petersson H, Johansson M, Persson N-K (2018) Quantifying shedding of synthetic fibers from textiles; a source of microplastics released into the environment. Environ Sci Pollution Res 25:1191–1199. https://doi.org/10.1007/s11356-017-0528-7
14. Zhang Q et al (2020) A review of microplastics in table salt, drinking water, and air: direct human exposure. Environ Sci Technol. https://doi.org/10.1021/acs.est.9b04535
15. Staff S (2020) For the first time, microplastics have been found in fresh fruits and vegetables; How does it affect us? https://swarajyamag.com/news-brief/for-the-first-time-microplastics-have-been-found-in-fresh-fruits-and-vegetables-how-does-it-affect-us
16. Imster E (2020) Plastic rain: more than 1,000 tons of microplastic rain onto western US. https://earthsky.org/earth/microplastic-rain-western-us
17. Brahney J, Hallerud M, Heim E, Hahnenberger M, Sukumaran S (2020) Plastic rain in protected areas of the United States. Science 368:1257–1260
18. Jönsson C, Arturin OL, Hanning AC, Landin R, Holmström E, Roos S (2018) Microplastics shedding from textiles—developing analytical method for measurement of shed material representing release during domestic washing. Sustainability 10:2457. https://doi.org/10.3390/su10072457
19. Hearle JWS, Lomas B, Cookie WD (1998) Atlas of fiber fracture and damage to textiles. Woodhead Publishing Ltd, London
20. O'Loughlin C (2018) Fashion and microplastic pollution, investigating microplastics from laundry. Ocean Remedy. https://cdn.shopify.com/s/files/1/0017/1412/6966/files/Fashion_and_Microplastics_Ocean_Remedy_2018.pdf. Accessed 15 Ap 2020
21. Boucher J, Friot D (2017) Primary microplastics in the oceans: a global evaluation of sources. IUCN, Gland, Switzerland, 43 pp. https://doi.org/10.2305/IUCN.CH.2017.01.en
22. Magnusson K, Norén F (2014) Screening of microplastic particles in and down-stream a wastewater treatment plant. Swedish Environmental Protection Agency
23. Zubris KAV, Richards BK (2005) Synthetic fibers as an indicator of land application of sludge. Environ Pollut (Barking, Essex: 1987) 138:201–211
24. De Falco F, Di Pace E, Cocca M, Avella M (2019) The contribution of washing processes of synthetic clothes to microplastic pollution. Sci Rep 9:6633. https://doi.org/10.1038/s41598-019-43023-x
25. Napper IE, Thompson RC (2016) Release of synthetic microplastic plastic fibres from domestic washing machines: effects of fabric type and washing conditions. Mar Pollut Bull 112(1–2):39–45. https://doi.org/10.1016/j.marpolbul.2016.09.025
26. Pirc U, Vidma M, Mozer A, Kržan A (2016) Emissions of microplastic fibers from microfiber fleece during domestic washing. Environ Sci Pollution Res 23:22206–22211. https://doi.org/10.1007/s11356-016-7703-0
27. Bishop D (1995) Physical and chemical effects of domestic laundering processes. In: Carr CM (ed) Chemistry of the Textiles Industry. Springer, New York, pp 125–172
28. Astrom L (2016) Shedding of synthetic microfibers from textiles. Gothenburg University. https://bioenv.gu.se/digitalAssets/1568/1568686_linn—str–m.pdf. Accessed 15 Apr 2020
29. De Falco F, Pia Gullo M, Gentile G, Di Pace E, Cocca M, Gelabert L, Brouta-Agnesa M, Rovira A, Escudero R, Villalba R, Mossotti R, Montarsolo A, Gavignano S, Tonin C, Avella M (2018) Evaluation of microplastic release caused by textile washing processes of synthetic fabrics. Environ Pollution 236:619–925. https://doi.org/10.1016/j.envpol.2017.10.057
30. Hernandez E, Nowack B, Mitrano DM (2017) Synthetic textiles as a source of microplastics from households: a mechanistic study to understand microfiber release during washing. Environ Sci Technol 51(12):7036–7046. https://doi.org/10.1021/acs.est.7b01750

31. Cesa FS, Turra A, Herminio Checon H, Leonardi B, Baruque-Ramos J (2019) Laundering and textile parameters influence fibers release in household washings. Environ Pollution 257:113553. https://doi.org/10.1016/j.envpol.2019.113553
32. Yang L, Qiao F, Lei K, Huiqin Li Yu, Kang SC, An L (2019) Microfiber release from different fabrics during washing. Environ Pollut 249:136–143. https://doi.org/10.1016/j.envpol.2019.03.011
33. Zambrano MC, Pawlak JJ, Daystar J, Ankeny M, Cheng JJ, Venditti RA (2019) Microfibers generated from the laundering of cotton, rayon and polyester based fabrics and their aquatic biodegradation. Marine Pollution Bull. 142:394–407. https://doi.org/10.1016/j.marpolbul.2019.02.062
34. Lant NJ, Hayward AS, Peththawadu MMD, Sheridan KJ, Dean JR (2020) Microfiber release from real soiled consumer laundry and the impact of fabric care products and washing conditions. PLoS One 15(6):e0233332. https://doi.org/10.1371/journal.pone.0233332
35. Kelly RM, Lant NJ, Kurr M, Grant Burgess J (2019) Importance of water-volume on the release of microplastic fibers from laundry. Environ Sci Technol 53:11735–11744. https://doi.org/10.1021/acs.est.9b03022
36. Deroberts N (2019) Washing laundry may be an underappreciated source of microplastic pollution. https://blogs.ei.columbia.edu/2019/08/22/laundry-microplastic-pollution/
37. Cotton L, Hayward AS, Lant NJ, Blackburn RS (2020) Improved garment longevity and reduced microfibre release are important sustainability benefits of laundering in colder and quicker washing machine cycles. Dyes Pigm. https://doi.org/10.1016/j.dyepig.2019.108120
38. O'Brien S, Okoffo ED, O'Brien JW et al (2018) Airborne emissions of microplastic fibres from domestic laundry dryers. Sci Total Environ. https://doi.org/10.1016/j.scitotenv.2020.141175
39. Francisco Belzagui M, Crespi M, Alvarez A, Gutierrez-Bouzan C, Vilaseca M (2019) Microplastics' emissions: microfibers' detachment from textile garments. Environ Pollution 248:1028–1035. https://doi.org/10.1016/j.envpol.2019.02.059
40. O'Loughlin C (2018) Fashion and microplastic pollution, investigating microplastics from laundry. Ocean Remedy. https://cdn.shopify.com/s/files/1/0017/1412/6966/files/Fashion_and_Microplastics_Ocean_Remedy_2018.pdf
41. Cai Y, Yang T, Mitrano DM, Heuberger M, Hufenus R, Nowack B (2020) Systematic study of microplastic fiber release from 12 different polyester textiles during washing. Environ Sci Technol 54(8):4847–4855. https://doi.org/10.1021/acs.est.9b07395
42. Markus Sillanpää and Pirjo Sainio (2017) Release of polyester and cotton fibers from textiles in machine washings. Environ Sci Pollution Res 24(23):19313–19321. https://doi.org/10.1007/s11356-017-9621-1
43. Putchta R (1984) Cationic surfactants in laundry detergents and laundry aftertreatment aids. J Am Oil Chemists' Soc 61(2):p367
44. Chiweshe A, Crews P (2000) Influence of household fabric softeners and laundry enzymes on pilling ratings and breaking strength. Text Chem Color Am Dyest Rep 9:41–47
45. Hartline N, Bruce N, Karba S, Ruff E, Sonar S, Holden P (2016) Microfiber masses recovered from conventional machine washing of new or aged garments. Environ Sci Technol 50(21):11532–11538. https://doi.org/10.1021/acs.est.6b03045

Microplastics in Dentistry—A Review

T. Chandran, Unnikrishnan Vishnu, and A. K. Warrier

Abstract Microplastics in the environment have become a public health concern over the past few years. Items of toothpaste and composite restorative materials are the primary dental products that contribute to the microplastic pollution of the environment. In terms of dental origin, toothpaste containing plastic particles <5 mm, form the source of primary microplastics. Secondary microplastics are formed from the resin-based composite restorative materials which degrade within the oral cavity or may be released during the process of finishing and polishing of restorations. At the same time, there is minimal awareness amongst the population regarding the use of microplastics in personal-care products. The prolonged use of toothpaste containing microbeads can cause abrasion of tooth enamel and dentine. The microbeads can get entrapped in the gingival sulcus leading to gingivitis and periodontitis. The resin-based composites used as direct restorative materials, pit and fissure sealants and in CADCAM milling can release monomers like Bisphenol A-glycidyl methacrylate (Bis-GMA) which is a potent environmental hazard as it gets dumped in the landfill. Studies have confirmed the damage to aquatic life caused by microplastics which can lead to the extinction of various aquatic species in future. They also get adsorbed to biotoxins and through the process of bioaccumulation enter the food chain. A few countries have enacted legislation which limits the use of microplastics in health care products. However, many do not impose such strict regulations. Therefore, an educational and regulatory approach toward the use of microplastics is mandatory to control the emerging threat of microplastics to the environment. This chapter is intended to create awareness among the public about the hazardous effects of microplastics in dental care products and promote insights into policymaking.

T. Chandran
Nitte (Deemed to be University), AB Shetty Memorial Institute of Dental Sciences (ABSMIDS), Department of Public Health Dentistry, Mangalore, India

U. Vishnu (✉) · A. K. Warrier
Manipal Institute of Technology, Manipal Academy of Higher Education, Manipal 576104, India
e-mail: vishnu.u@manipal.edu

A. K. Warrier
Centre for Climate Studies, Manipal Academy of Higher Education, Manipal 576104, India

S. S. Muthu (ed.), *Microplastic Pollution*, Sustainable Textiles: Production, Processing, Manufacturing & Chemistry, https://doi.org/10.1007/978-981-16-0297-9_6

Keywords Microplastics · Dentistry · BPA · Denture base materials · Toothpaste · Resin-based composite · Environmental pollution · Biotoxins

List of Abbreviations

PEMRG	PlasticsEurope's Market Research and Statistics Group
BPA	Bisphenol A
RBC	Resin-based composites
HEMA	Hydroxy ethyl methacrylate
TEGDMA	Triethylene glycol dimethacrylatef
UDMA	Urethane-dimethacrylate
PMMA	Poly Methyl Methacrykate
EGDM	Ethylene glycol dimethacrylate agent
BIS GMA	Bisphenol A-glycidyl methacrylate
BIS DMA	Bisphenol A dimethacrylate
WWTP	Waste water treatment plant
MP	Microplastic
PE	Polyethylene
POP	Persistent Organic Pollutants
PCB	Polychlorinated biphenyl
MMA	Methyl methacrylate
UV	Ultra Violet

1 Introduction

Plastic by its general application is a group of synthetic polymers derived from fossil fuel or biomass-derived plastics. At the current production rate and usage trends, it is estimated to reach a production of 2000 million tonnes by 2050 [44]. With the global production of 1.5 million t in 1950 to around 360 million t of production worldwide in 2018 (PEMRG), plastic demand is increasing by each day due to its adaptability to varied industries and societal needs [1]. The same benefits of the material like its durability, resistance to degradation, etc., has made it an indispensable part of the human life and also added to the woes in its waste management. Around 4.8–12.7 million tonnes of plastic enter our ocean every year [30].

The small size plastics popularly called as microplastics is the subject of concern these days as they can affect the environment and human health in various ways. There is no clear-cut universal guideline in terms of defining microplastic. However, the particles with size <5 mm are generally considered as microplastics [2, 20]. They are further classified based on origin into primary and secondary microplastics. Plastics originally manufactured as small size commonly found in textile fabrics, drug, and cosmetic products are called primary microplastics [9, 14]. Degradation or fragmentation of larger plastic debris, by UV-radiation and mechanical abrasion results in the formation of secondary microplastics [57]. Microplastics thus released to the environment affects the environment due to the release of chemicals present in them either in the form of additives originating from plastics itself or as chemicals absorbed from surrounding.

Additives give the desired qualities of color and transparency and also resist degradation from ozone, temperature, bacteria, and thermal and electrical resistance [26]. Many of these chemicals have negative effects on the health of human as they have proven to be endocrine disruptors like Bisphenol A (BPA), phthalates, as well as some of the brominated flame retardants [12]. Endocrine disrupting chemicals (EDC) can alter the homeostasis of the endocrine system. They can antagonize the action of natural hormones, alter the pattern of synthesis [13, 41].

The exposure of these chemicals in the field of dentistry is of great concern with the presence of them in the dental restoration composites as resin composites have become a viable alternative to the dental amalgam. Composites are more advantageous than amalgam due to their aesthetic and desirable methods of restoration. These features often makes it preferable and outweigh the local risks associated [4]. The typical application of polymers in the field of dentistry relates to the use in dentures, fillers, and impression materials.

Transmission of harmful components from these composite resins happens during clinical application also. These particulates containing part-polymerised monomer are released into waste water after their polishing and finishing process. These particulate matter finally reach the water bodies. There is an increase in the usage of highly polymerised RBC for the manufacture of dental crowns, inlays, and onlays which create fine micro-particle waste powder in large volumes during the milling process, and gets released into municipal wastewater. These all add to the woes of the microparticulate and particularly microplastic pollution in environment.

European food safety authority (EFSA) has revised the daily tolerable limit of BPA in food intake to a temporary level of 4 μg/kg bw/day which is far lower than the limits they have prescribed earlier. Until 2015 the limit was put at 50 μg/kg bw/day. There is significant uncertainty exists among the practitioners and scientists regarding the cytotoxicity of these chemicals on human body. Based on the studies conducted it is worthwhile for the clinician to understand the potential toxicity of the chemicals they administer on the patients and the protective procedures that they can adopt.

There is a great dearth and uncertainty in terms of the studies related to polymer exposure from dental materials [36] among practitioners. Though significant number of cross-sectional studies on knowledge on hazardous and general waste among practitioners is conducted, it would be worthwhile to have a similar study on another potential area of toxicity like polymer exposure from dental materials. A summary of the current studies being done in the field of dentistry related to microplastic usage is listed in Table 1. This chapter would address the current studies being done in

Table 1 Summary of current studies being done on the microplastic use in dentistry

Sl. no	Objective of study	Methodology adopted	Findings	References
1	Quantitative characterization of the bisphenol-A (BPA) released from orthodontic adhesives after artificial accelerated aging	In vitro assessment of the release of BPA after various time intervals of 1 day, and weekly intervals of 1,3 and 5 weeks from chemically cured, no-mix adhesive and a visible light-cured adhesive bonding agents	BPA release from light-cured or chemically cured, no-mix adhesives did not reach the 0.1 ppm level	Eliades et al. [18]
2	To assess the bisphenol A (BPA) released from an orthodontic adhesive with various light curing tip distances	In vitro assessment of BPA released from orthodontic adhesive at light curing tip distances of 0, 5 and 10 mm using Fourier transform infrared spectroscopy	BPA release was greater in specimens cured with a greater light-curing tip distance	Sunitha et al. [49]
3	Comparison of estrogenicity of heat cured and light cured dental adhesives which were collected during the debonding process of fixed orthodontic appliances	In vitro assessment of the influence of eluents from the solution of two type of adhesives to estrogen responsive cell line	BPA in self cured resins increased the proliferation of MCF-7 breast cancer cells by 160% while there was a 128% increase by BPA released from light cured resin adhesives	Gioka et al. [23]
4	Identification and quantification of microplastics in tooth paste	Particles were quantified and then characterized by microscopic evaluation and surface chemistry analysis	4 out of 20 samples showed the presence of polyethylene type MPs in the concentration range of 0.4–1%	Ustabasi and Baysal [55]
5	Identification and quantification of polyolefins in tooth paste	Quantitation with high-temperature gel-permeation chromatography	Identified the presence of 0.17% of polyethylene type MPs in a single sample they tested	Hintersteiner et al. [28]

(continued)

the field of dentistry to assess the use of microplastics in dentistry, toxicity level of various composite resins being used and the impact of them on environment and human.

Table 1 (continued)

Sl. no	Objective of study	Methodology adopted	Findings	References
6	Assessment of salivary and urinary concentration of BPA after RBC placement	In vivo study on urine samples followed by analysis using Liquid chromatography/mass spectrometry	BPA levels in urine levels following sealant placement, have shown an increase in urine BPA for upto 30 h	Kingman et al. [31]
7	Identification and characterization of microplastics in toothpaste and personal care products	Identification using particle analysis software image J 1.51 and composition by FTIR	Found LDPE type MPs in the concentration of 7% in the sample tested	Praveena et al. [42]
8	Assessment of weathering impact and uptake of polyethylene microplastics from toothpaste in mussels	Mussels where exposed to Polyethylene (PE) extracted from toothpaste	100 ml of a popular toothpaste brand and extracted 100 mg of polyethylene MPs PE particle ingestion resulted in structural changes to the gills and digestive gland, as well as necrosis in other tissues such as the mantle	Bråte et al. [7]
9	Assessment of interaction of microplastics derived from toothpaste with four types of bacteria	Study on Microplastics extracted from commercially available toothpaste samples interacted with four types of bacteria under both standard and seawater conditions	Biochemical responses were similar in both media, the difference between the cell wall and microplastics surface charge was important only in seawater	Ustabasi and Baysal [56]
10	To identify the relation between usage of dental amalgam and psychosocial problems	Conducted a randomized trial among children of age group 6–10 years	Study has shown a reflection of the harmful effects of resin based composites as they had shown detrimental effects to the children behavior compared to amalgam group	Bellinger et al. [5]

(continued)

Table 1 (continued)

Sl. no	Objective of study	Methodology adopted	Findings	References
11	Association between dental sealants or restorations and urinary bisphenol A levels in schoolagedchildren in the United States	Cross sectional study Children were grouped based on number of sealants present: 0, 1–3, 4–6, and 7–16	Compared to children with no sealants, children with 7–16 sealants had 25% higher mean urinary BPA levels	Martin and Derouen [37]
12	Assessing BPA derivatives in composite resins marketed in Europe	Composition was determined from both material safety data sheets and a standardized questionnaire sent to manufacturers	Study showed out of the 160 composite resins studied except 18 of them all the other had derivatives of BPA in them	Dursun et al. [16]
13	To evaluate the knowledge, attitude, and behavior of Restorative, Orthodontic, and Pediatric Dentistry Departments' members at King Abdulaziz University (KAU), Jeddah, Saudi Arabia, toward bisphenol A (BPA) dental exposure	Self-administered questionnaire was distributed among all members (182 members) at KAU in Jeddah, Saudi Arabia	Most of the participants have not checked the BPA content (99.1%) of the dental materials used in restoration practices so far very few have knowledge about the development in the field (11%) Only 9% followed the recommended guidelines in dental practices to reduce exposure to BPA	Bagher et al. [3]
14	Assessing BPA concentration in saliva after composite restoration	Tested salivary BPA concentrations after restoration treatment with monomer based composite resins using ELISA system	Results showed several tens to 100 ng/ml BPA in the saliva after filling the cavities	Sasaki et al. [47]
15	Evaluation of cellular toxicity of acrylic resin in comparison with heat cure resin	Acrylic resin placed in 24-well culture plates along with L929 mouse fibroblast cell line and cytotoxicity was assessed by MTT assay and ELISA	Cytotoxicity was shown to be higher in the auto polymerised acrylic resin compared to the heat cured resin	Saravi et al. [46]

(continued)

Table 1 (continued)

Sl. no	Objective of study	Methodology adopted	Findings	References
16	Assessing the impact of ultrasonic treatment he amount of residual methyl methacrylate monomer in one heat-polymerized acrylic resin	Residual monomer was extracted and analyzed using high performance liquid chromatography	Ultrasonic treatment could enhance the extraction rate of the residual monomer from the resin and could cause postpolymerization of the residual monomer	Charasseangpaisarn and Wiwatwarrapan [11]

2 Dental Materials and Microplastics

Microplastics are generally defined by the higher polymer content, their peculiar properties of being insoluble in water other than the size criteria of being 5 mm. The degradability of the polymer material also need to be taken into account in case of deciding the nature of microplastics. Due to this insolubility they have the ability to persist in the aquatic environment.

The various composite resins used in dentistry include these polymer content like monomethyl methacrylate, epoxy resins. HEMA, TEGDMA, UDMA, etc. The denture polymer resins replace the lost or damaged dental tissues and play a functional and morphological role in the oral cavity. There are a variety of uses for resins in dentistry. They are used in the form of

a. Denture bases, artificial teeth, denture base liners
b. Impression materials
c. Resin-based dental cements
d. Composite filling materials
e. Pit and fissure sealants
f. Obturating materials.

2.1 Poly Methyl Methacrylate Based Denture Base Materials

The denture base polymer can be heat cured polymers, auto-polymerized polymers, light activated or microwave activated resins, and thermoplastic resins. The acrylic denture bases is made from the polymer Poly Methyl Methacrykate (PMMA) in the powder form, and the liquid monomer Mono Methyl Methacrylate, and cross linking agent Ethylene glycol dimethacrylate agent (EGDM). These biomaterials induce a biological tissue response. These materials should be biocompatible and when used in a living environment should not cause any adverse side effects. The biocompatibility is a dynamic process as both the materials and host tissues undergo changes over time. The acrylic resins used in dentistry like any other dental biomaterial not fully inert and can occasionally cause local or systemic toxicity. All dental products should be certified by International Organization for Standardization for its safe use [33].

The monomer methyl methacrylate is supplied as liquid and it is mixed with the polymer which is in powder form. When mixed, the polymer gets partially dissolved in the monomer to form a plastic dough. This dough is packed into the moulds and it gets polymerized. This polymer has very good stability, doesn't undergo discolouration easily, has good ageing properties and hence is widely used in dentistry.

The polymerization of acrylic monomers is often incomplete. The rate of polymerization largely depends on the mode of polymerization. Heat cured products often leave less unpolymerized resin than self-cured polymerization. The residual monomers left leached out of the dental products. They dissolve in saliva and get absorbed into the oral mucosa, skin and can also reach the gastrointestinal system. The health hazards caused by these materials are often due to their cytotoxicity. Residual MMA gets hydrolysed and forms methacrylic acid which is a proved allergen and tissue irritant. EGDM is identified as the strongest allergen in acrylic materials [27].

To reduce the amount of residual polymers released polymerization at high-temperature close to the glass transition temperature of the materials used is recommended. Microwave and light-curing releases less volume of residual monomers but their use in denture bases is complicated and expensive. The residual monomer release should be in the range of 1–3%. Microwave post-polymerization as well as immersing the dentures in a water bath for about 1–7 days before inserting into the oral cavity can significantly decrease the release of residual monomers [54].

The localized manifestation of the effects of acrylics include stomatitis, cheilitis, candidiasis, and painful sensations, diffused erythema and urticaria. The allergy to the acrylics can also be in an extensive form as in Erythema multiforme. Most of the cases report an acute form of irritation to dental resins and they subside once the irritant is removed. In contrast chronic denture wearers who are elderly people report chronic form of tissue irritation as in fibrous hyperplasia [19]. Orthodontic patients and prosthesis wearers gradually ingest the residual monomers and this require special attention in the context of systemic toxicity though systemic toxicity is rarely reported [24].

Occupational hazards associated with the use of dental monomers and polymers is a matter of major concern. Dental technicians often fall prey to this. In order to achieve greater precision, they often avoid the use of gloves. This can cause contact dermatitis and eczema of skin of distal phalange and palmar surface of finger tips. The affected parts experience dryness, itching, cracking, peeling, and swelling. The inhalation of the polymer can cause asthma, anorexia, decreased gastric motor activity, headache, and drowsiness. In severe cases can also cause neurological dysfunctions like paraesthesia and neuropathy [34]. Hence dentists and auxiliaries should avoid direct contact with the polymers by practicing no touch techniques, avoid inhalation by using protective masks while working with these materials and ensure that the operatory is well ventilated.

2.2 *BIS Phenol A from Orthodontic Appliances*

In orthodontic appliances, there is an extensive use of polymers as polyurethanes for elastomeric ligatures and chains, polycarbonates for esthetic brackets, polyamide-based wire sleeves, lip bumper appliances made of poly propylene. Bis Phenol A is often used in the manufacture of dental resins as a precursor of BIS GMA or bis phenol A Dimethacrylate. Sudies have shown that this BPA is a culprit in endocrine dysfunctions causing peripubertal mammary gland development in mice, male feminization, and early puberty of females. Systemic intake of BPA can be through ingestion, inhalation, and through the dental pulp [32].

Most of the times, the orthodontic appliances are worn for an average duration of two years and the retainers are worn for an extended period. In the case of bonded lingual retainers, a larger surface of the adhesive used for bonding is exposed to the oral cavity and the thickness of the resin adhesive promotes incomplete polymerization. The leaching of materials from this resin can initiate adverse health effects. Clinicians are encouraged to do a prophylaxis with pumice over the surface of these resin adhesives to remove the uncured material [17]. Debonding of the brackets is a major cause of exposure to resins for the dental personnel.

In vitro studies on residual TEGDMA released from adhesives have shown that it can cause chromosomal anomalies by deleting DNA chain sequence. Also it was observed that this resin adhesives mimic estrogenic activity after stimulated debonding. Polycarbonate brackets degrade in the oral cavity and release BPA. It was noticed that the polycarbonate base ceramic brackets release more amount of umpolymerized resins than the ceramic brackets [29].

The orthodontic adhesives are exposed to the oral environment through the peripheral margins of bracket, through the fixed maxillary and mandibular retainers and while removing the fixed appliances which involves removal of the adhesive by grinding with burs [23]. The release of polymers from margins is less compared to that from the fixed retainers which remain on the teeth either for a long duration or entire lifetime [43]. The particles released during removal of the appliances is a mixture of filler degradation products, polymer matrix pieces, and particles from the wear of bur used. This aerosol can cause health hazards to both the patient as well as the dentist. The Bis-GMA released can eventually lead to formation of BPA which is a known endocrine disruptor [50].

Various in vitro studies done to estimate the release of BPA from orthodontic adhesives show conflicting results. A study by [18] in an in vitro environment assessed the release of BPA after various time intervals of 1 day, and weekly intervals of 1, 3, and 5 weeks. The study did not show any evidence of BPA release. However, the authors mentioned that the results could not be extrapolated in an in vivo environment. Another study by [49] attempted to correlate the distance between the light-curing unit and the degree of conversion of the polymers. The study concluded that an inverse relation exists between the distance of the curing unit and the rate of polymerization. A reduction in the degree of polymerization was accompanied by an increased rate of polymers into the oral environment. The estrogenicity of light cured and chemically cured adhesive resins was compared by [18]. It was noted that the eluents containing particles of resins could not induce any proliferation in the active estrogen-sensitive MCF-7 breast cancer cells and the MB-231 adenocarcinoma cells which served as the control. Another study by [23] compared the estrogenicity of heat cured and light cured dental adhesives which were collected during the debonding process of fixed orthodontic appliances. It was observed that BPA in self-cured resins increased the proliferation of MCF-7 breast cancer cells by 160% while there was a 128% increase by BPA released from light cured resin adhesives.

Regarding the in vivo effects of BPA there has been a controversy in the safely tolerated levels. While the European Food Safety Authority has proposed a level of 50 μg/kg/day, study by [53] mentioned that levels lesser than 50 μg can alter biological activities of cells.

Reccomendations to decrease the exposure to residual monomers include keeping the light cure tip as close as possible to the adhesive resin, using indirect polymerization across the bracket edge to ensure minimal amount of unpolymerization, making the patient rinse the mouth in the first hour after placement of brackets or bonding the retainer and removing the unpolymerized layer by doing a pumice prophylaxis [32].

Evidences from in vivo studies are limited and more studies are needed to elaborate the effects of resins used in orthodontic appliances.

2.3 BIS Phenol A from Composites and Pit and Fissure Sealants

BPA by itself is not a component of sealants or composite restorations. It is the starting material of Bis-GMA, which is a common monomer used in composites. Hence it is present in trace amounts in these monomers. Another common source of release of BPA is through hydrolydid of Bis-DMA by salivary esterases. Bis-DMA is another commonly used monomer. Unlike Bis-DMA, Bis-GMA does not undergo hydrolysis by salivary esterases. On exposure to Oxygen, the superficial layer of the sealants remain unolymerized. This causes leaching of the unpolymerized monomer into the saliva. This leaching is maximum during the initial hours following the sealant placement which gradually decreases. Studies measuring the BPA levels in urine levels following sealant placement, have shown an increase in urine BPA for up to 30 h [31].

The resin-based composites are made of filler particles, matrix composed of resins, and coupling agents. The matrix is composed of monomers like UDMA, Bis-GMA, EGDMA, TEGDMA, etc.

2.4 Toothpastes and Microplastics

Ustabasi and Baysal [55] conducted a study on 20 toothpaste samples to identify and quantify the type of microparticles present. They found that 4 out of 20 samples showed the presence of polyethylene type MPs in the concentration range of 0.4–1%. Another study by [28] identified the presence of 0.17% of polyethylene type MPs in a single sample they tested. Praveena et al. [42] found polyethylene type MPs in the concentration of 7% in the sample tested. Bråte et al. [7] conducted a study on 100 ml of a popular toothpaste brand and extracted 100 mg of polyethylene MPs. In contrary to these studies, the studies by [35] reported the absence of MPs in all the samples tested.

Ustabasi and Baysal [56] attempted to find the effect of toothpaste derived MPs on bacterial strains found in sea water. The bacterial strains studied were *B. subtilis, S. aureus, P. aeruginosa,* and *E. coli.* The bacteria were studied under laboratory conditions and in sea water. In the standard condition, only *P. aeruginosa* was affected significantly with a growth inhibition of 20–35%. In sea water, *B.subtilis* was affected significantly with a growth inhibition of 20–30% while *S. aureus* showed a slight growth inhibition of 0–30%. This study also showed the discrepancy in results when studied under laboratory and actual marine environments.

Polyethylene is the most common MP worldwide and toothpaste contain up to 1.8% polyethylene [10]. Similar MPs were also found in effluent of waste water treatment plant [WWTP]. Though the WWTPs remove MPs efficiently, their concentration in effluents is still high [38]. MPs in seawater get coated with a bacterial biofilm called slime or conditioning film. They are also subjected to physical stresses, varying temperatures, UV-radiation, oxidation, and salinity. The weathered or conditioned MPs have an altered surface morphology and increased density and they sink to bottom. This makes their availability more for a variety of marine organisms [51]. Mussels are considered as indicators of microplastic pollution in marine environment. A study was conducted by Bråte et al. [7] to identify the effects of PE MPs on M. galloprovincialis. They observed that weathered PE microparticles were ingeted more compared to virgin PE particles. The various histopathological changes in the exposed mussels include a decrease or complete absence of contacts between filaments in the gills, haemocyte infiltration in gills, replacement of ciliated epithelium by squamous epithelium in the digestive glands. In organs like mantles and gonads severe tissue necrosis was noted.

3 Polymer Degradation and Impacts

3.1 Saliva Components

Water forms the largest proportion of saliva. RBC being a polar material, water molecules percolate into the polymer network causing diffusion of uncured monomers and additives from the network. Thee polymers can be degraded by passive hydrolysis and enzymatic degradation. The extent of enzymatic degradation depends on the extend of cure of the resin. The ester groups are available for reaction in a more loosely cross-linked network [22]. Most importantly the composition of monomers determines the extent of degradation.

3.2 Masticatory Forces

In the oral environment the composite materials are subjected to relatively low repetitive masticatory or chewing forces. This continuous mechanical loads eventually lead to degradation and initiation of cracks within the restoration. This process is further accelerated by residual stresses and pre-existing voids induced during the processing stage of the material [15].

3.3 Thermal and Chemical Changes

The inta oral temperature gradient caused by ingestion of different kinds of food produces an unfavorable environment for the RBC. The coefficient of thermal expansion of these materials is not as same as the natural teeth. This in turn can induce surface stresses leading to degradation of RBC. The changes in PH caused by the varieties of food and beverages can affect the dental materials directly [6].

3.4 Oral Microbes

In vitro studies have shown that the presence of bacteria on the surface and its interaction with the polymer can cause degradation and surface roughness of the composite restorations [25].

4 Environmental Impact

At the point of origin, disposal of the waste material during the manufacturing of resin-based composite is the first pollution event. This manufacturing waste is used for landfill after polymerization, the verification of which is rather difficult. In dental clinics, the excess dental composites after treatment as well as expired materials in syringes is considered as municipal solid waste and is discarded in landfill sites. This leads to a reaction between the landfill leachate and RBC leading to the release of various components [40]. Landfill leachate is formed by the bacterial and fungal action in the contents of landfill which causes its decomposition. The physical parameters like pH, temperature, and oxygen content of the landfill leachate varies over a period of time which affects its reactivity. Often a heterogenous mixture of commercial, mixed industrial, and municipal waste are dumped into a landfill site giving rise to leachate comprising of organic and inorganic matter, xenobiotic compounds, and heavy metal ions. This reactive leachate interacts with RBC which releases its constituent monomers, oligomers, and BPA. The United States Environmental Pollution Agency (USEPA) Office through computer simulations proposed that an accidental release of dental composites while transportation of dental waste or any malfunction of the landfill liners are the only potential causes of environmental contamination by dental composites [39].

Another reason for concern is when RBCs are discarded in coastal landfill sites and areas that may be subjected to coastal erosion. If such sites are subjected to natural disasters like floods, or lost to sea due to coastal erosion, RBCs can cause environmental pollution. In view of this, if RBCs are subjected to incineration, the risk can be mitigated. In clinical practice, the finishing and polishing of composite restorations while placement as well as while removing a damaged composite restoration, there is release of particulate as well as microparticles consisting of monomers which are released into the waste water [8].

Onlays, inlays, and crowns are made of highly polymerized RBC through CADCAM milling. This process releases fine microparticles which are released into municipal waste water systems. These microparticles which are <5 mm act as direct pollutants and can also attach to biotoxins or Persistent Organic Pollutants (POPs) like polychlorinated biphenyl (PCB). These microplastics after adsorbing POPs can gain access to food chain by bioaccumulation. They are ingested by plankton, fish, mussels, barnacles, and seabirds. The harmful effects in aquatic life caused by consumption of microparticles can be due to accumulation of such particles in the organism and toxicity arising from them, contaminants leaching from them and ingestion of other attracted pollutants which are bound to these particles. Additives released from these microplastics like BPA can disrupt the endocrine function of aquatic organisms affecting their development, reproduction, and motility [39]. BPA even at small concentrations of ng/L to mg/L is seen to affect the endocrine function of crustaceans and fishes and also causes molecular as well as whole body effects.

5 Impact on Human

Studies have shown that the release of free monomer continues for months after a restoration is placed. They are detected in saliva and urine of the individuals. This free monomer reaches the environment through human excreta. When individuals with RBCs are cremated or interred, the RBC monomers from the oral cavity reach the environment through crematoria waste and emissions or by exposure to ground water. An RBC restoration that has been inside the oral cavity of an individual for years releases lesser monomer compared to a relatively newer restoration. Compared to the conventional RBCs, the CADCAM milled RBCs are highly polymerized and consequently released lesser monomer [31].

Study by Bellinger et al. [5] showed worse psychosocial behavior among children who had done dental restorations with composites than with the mercury-laden amalgam. They conducted a randomized trial among children of age group 6–10 years. Their directional hypothesis of the usage of dental amalgam in psychosocial problems was not confirmed, however, the study has shown a reflection of the harmful effects of resin-based composites as they had shown detrimental effects to the children behavior compared to amalgam group. This is in confirmation with the toxicity of resin monomers [48]. They showed the importance of concentrating more into the dental-based composites used as replacement of amalgam for restoration.

Martin et al. [37] studied the possible association of dental sealants and Urinary BPA levels in children. The number of occlusal-surface sealants observed in children of different age group were observed. Though statistically not significant, study showed higher BPA concentration with more number of sealants. However, the study had the limitation of not collecting the data regarding the dental material and its composition which is very much important in such a study. Different products have different resin leaching properties depending on the manufacturers [52]. Dursun et al. [16] has done a study to assess the BPA or its derivative present in the composite resins marketed in Europe. The study showed out of the 160 composite resins studied except 18 of them all the other had derivatives of BPA in them. Many manufacturers (25.8%) seem to be reluctant in disclosing the exact composition of the products which makes it difficult for the practitioner to choose the ideal one. The study pointed out the importance of concentrating more into the alternatives for resin materials like inorganic biomaterials, ceramic, and carbomers.

Bagher et al. [3] conducted a study among the restorative, peadiatric, and orthodontic department members to assess the knowledge toward the exposure of the BPA chemicals or derivatives. Most of the participants have not checked the BPA content (99.1%) of the dental materials used in restoration practices so far and very few have knowledge about the development in the field (11%). Also only 9% followed the recommended guidelines in dental practices to reduce exposure to BPA. Sasaki et al. [47] tested salivary BPA concentrations after restoration treatment with monomer-based composite resins. Results showed several tens to 100 ng/ml BPA in the saliva after filling the cavities. However, they showed depending on the concentration of monomer used the BPA content can be significantly brought down to less than 10 ng/l by proper gargling after the procedure. This particularly important in the case of the vulnerable group like pregnant women and children after they undergo the treatment.

Sasaki et al. [46] studied the cellular toxicity of auto polymerised acrylic resin in comparison to the heat cured resin. Cytotoxicity was shown to be higher in the auto polymerised acrylic resin compared to the heat cured resin. Cytotoxicity level however was not significantly different after a period of one week though it had a higher effect in the first 24 h. They stated the importance of immersing the denture materials in water during the first 24 h before administering it to the patient.

Charasseangpaisarn and Wiwatwarrapan [11] studied the interaction between ultrasonic treatment and acrylic resins and found significant reduction in the presence of residual monomer contents. MMA under heat-polymerization reduced the residual monomer content. It was found that ultrasonic treatment could enhance the extraction rate of the residual monomer from the resin and could cause postpolymerization of the residual monomer.

Sakr and Alhablain [45] studied the knowledge and attitude of dental students and intern practitioners about usage of nanotechnology in dental practice. Nano materials are widely used these days for the production of dental materials, however, they have reported cytotoxic effects particularly in studies done on animals [21]. Study showed that majority of the participant had less knowledge about nanotechnology in dentistry.

6 Recommendations for Future Studies

Though there are many studies done suggesting the potential toxicity of microplastics used in dentistry, most of them are in vitro assessments. More in vivo studies need to be done to assess the long-term impact of these polymers on human body. Majority of the research done regarding the impact of polymer resin and additives in dentistry has stated the harmful effects of these chemicals on the human body. However, there is a dearth of knowledge among the practitioners in general regarding its ill effects. There are limited study done on the knowledge and attitude of practitioners related to this subject though there are plenty of cross-sectional studies regarding general waste and hazardous waste impact. Sustainable use of these polymers in the field of dentistry requires strict adherence to policies which need to be taken care by the regulatory bodies. More insight into policy regulations with respect to manufacturing and treatment stages of RBC or polymer materials in dentistry is required in the present scenario.

7 Conclusion

The impact of microplastic polymer particles and there potential toxicity in dentistry is always a topic of debate. Since dentist are the connecting link between manufacturers of these dental materials and the patients they can do significant contribution in ensuring the right use of it. In order to prevent the harmful effects associated with the usage of these materials, monomer polymer conversion rate is crucial. Heat-cure resins should be preferred over self-cure resins.

Studies have shown that dental products containing the Bis-GMA have less estrogenicity than those containing bis-DMA. It is always better to store such dentures, and polymeric devices, in water under a temperature of 37 °C to remove the leachable material present. Also strict regulation on the usage of BPA need to be communicated to the manufacturers in curbing the over dosage of the same. It is also imperative for them to disclose the composition of the products to the customers and practitioners to ensure safe usage. Still more research need to be done to identify the long-term potential toxicity of these materials on the human body. Research into alternate materials like carbomers or biomaterials should be promoted to have a sustainable use of materials in dentistry.

References

1. Andrady AL, Neal MA (2009) Applications and societal benefits of plastics Applications and societal benefits of plastics. Philos Trans R Soc B 364:1977–1984. https://doi.org/10.1098/rstb.2008.0304
2. Arthur C, Baker JE, Bamford H (eds) (2009) Proceedings of the international research workshop on the occurrence, effects, and fate of microplastic marine Debris. NOAA Techn Memorandum NOS-OR&R-30
3. Bagher SM, Sabbagh HJ, Aldajani M, Al-Ghamdi N, Zaatari G (2019) Knowledge, Attitude, and Behavior of Restorative, Orthodontic, and Pediatric Departments' Members toward Bisphenol A Dental Exposures. J Int Soc Prevent Commun Dentistry 9(1):83
4. Bakopoulou A, Papadopoulos T, Garefis P (2009) Molecular toxicology of substances released from resin-based dental restorative materials. Int J Mol Sci 10:3861–3899
5. Bellinger DC, Trachtenberg F, McKinlay S, Zhang A, Tavares M (2008) Dental Amalgam and psychosocial status: the new England children's amalgam trial. J Dent Res 87(5):470–474
6. Bettencourt AF, Neves CB, de Almeida MS, Pinheiro, LM, e Oliveira SA, Lopes LP, Castro MF (2010). Biodegradation of acrylic based resins: a review. Dental Mater 26(5):e171–e180
7. Bråte ILN, Blázquez M, Brooks SJ, Thomas KV (2018) Weathering impacts the uptake of polyethylene microparticles from toothpaste in Mediterranean mussels (M. galloprovincialis). Sci Total Environ 626:1310–1318
8. Brand JH (2017) Assessing the risk of pollution from historic coastal landfills (Doctoral dissertation)
9. Browne MA (2015) Sources and pathways of microplastics to habitats. In: Bergmann M, Gutow L, Klages M (eds) Marine anthropogenic litter. Springer, pp 229–244. https://doi.org/10.1007/978-3-319-16510-3
10. Carr SA, Liu J, Tesoro AG (2016) Transport and fate of microplastic particles in wastewater treatment plants. Water Res 91:174–182
11. Charasseangpaisarn T, Wiwatwarrapan C, Leklerssiriwong N (2016) Ultrasonic Cleaning reduces the residual monomer in acrylic resins. J Dental Sci 11:443–448
12. Cingotti N, Jensen GK (2019) Health and Environment Alliance (HEAL). Food contact materials and chemical contamination. Health and Environment Alliance, Brussels, Belgium
13. Colborn T, Clement C (1992) Chemically-induced alterations in sexual and functional development: thewildlife/human connection, vol 21. Scientific. Pub. Co., Princeton, NJ, USA, p 403
14. Cole M, Lindeque P, Halsband C, Galloway TS (2011) Microplastics as contaminants in the marine environment: a review. Marine Pollution Bull 62(12):2588–2597
15. Drummond JL (2008) Degradation, fatigue, and failure of resin dental composite materials. J Dent Res 87(8):710–719
16. Dursun E, Fron-Chabouis H, Attal J-P, Raskin A (2016) Bisphenol A release: survey of the composition of dental composite resins bisphenol a release: survey of the composition of dental composite. Open Dentistry J 10(September):446–453
17. Eliades T, Viazis AD, Eliades G (1991) Bonding of ceramic brackets to enamel: morphologic and structural considerations. Am J Orthod Dentofac Orthop 99(4):369–375

18. Eliades T, Hiskia A, Eliades G, Athanasiou AE (2007) Assessment of bisphenol-A release from orthodontic adhesives. Am J Orthod Dentofacial Orthop 131:72–75
19. Ergun G, Mutlu-Sagesen L, Karaoglu T, Dogan A (2001) Cytotoxicity of provisional crown and bridge restoration materials: an in vitro study. J Oral Sci 43(2):123–128
20. Eriksen M, Lebreton LCM, Carson HS, Thiel M, Moore CJ, Borerro JC, … Ryan PG (2014) Plastic pollution in the world' s oceans: more than 5 trillion plastic pieces weighing over 250, 000 tons Afloat at Sea. PLoS One 9(12):1–15
21. Feng X, Chen A, Zhang Y, Wang J, Shao L, Wei L (2015) Application of dental nanomaterials: potential toxicity to the central nervous system. Int J Nanomed 10:3547–3565
22. Ferracane JL (2006) Hygroscopic and hydrolytic effects in dental polymer networks. Dent Mater 22(3):211–222
23. Gioka C, Eliades T, Zinelis S, Pratsinis H, Athanasiou AE, Eliades G, Kletsas D (2009) Characterization and in vitro estrogenicity of orthodontic adhesive particulates produced by simulated debonding. Dental Mater 25(3):376–382
24. Gonçalves TS, Morganti MA, Campos LC, Rizzatto SM, Menezes LM (2006) Allergy to autopolymerized acrylic resin in an orthodontic patient. Am J Orthod Dentofac Orthop 129(3):431–435
25. Gupta SK, Saxena P, Pant VA, Pant AB (2012) Release and toxicity of dental resin composite. Toxicol Int 19(3):225
26. Hahladakis NJ, Costas AV, Weber R, Iacovidou E, Purnell P (2018) An overview of chemical additives present in plastics: migration, release, fate and environmental impact during their use, disposal and recycling. J Hazard Mater 344:179–199
27. Hashimoto Y, Tanaka J, Suzuki K, Nakamura M (2007) Cytocompatibility of a tissue conditioner containing vinyl ester as a plasticizer. Dent Mater J 26(6):785–791
28. Hintersteiner I, Himmelsbach M, Buchberger WW (2015) Characterization and quantitation of polyolefin microplastics in personal-care products using high-temperature gel-permeation chromatography. Anal Bioanal Chem 407(4):1253–1259
29. Hougaard KS, Hannerz H, Feveile H, Bonde JP (2009) Increased incidence of infertility treatment among women working in the plastics industry. Reprod Toxicol 27(2):186–189
30. Jambeck JR, Geyer R, Wilcox C, Siegler TR, Perryman M, Andrady A, … Law KL (2015) Plastic Waste inputs from land into Ocean, vol 347
31. Kingman A, Hyman J, Masten SA, Jayaram B, Smith, C., Eichmiller, F.,… Dunn WJ (2012). Bisphenol A and other compounds in human saliva and urine associated with the placement of composite restorations. *Jthe Am Dental Assoc* 143(12):1292–1302
32. Kloukos D, Pandis N, Eliades T (2013) Bisphenol-A and residual monomer leaching from orthodontic adhesive resins and polycarbonate brackets: a systematic review. Am J Orthod Dentofac Orthop 143(4):S104–S112
33. Kostic M, Pejcic A, Igic M, Gligorijevic N (2017) Adverse reactions to denture resin materials. Eur Rev Med Pharmacol Sci 21(23):5298–5305
34. Leggat PA, Kedjarune U (2003) Toxicity of methyl methacrylate in dentistry. Int Dent J 53(3):126–131
35. Lei K, Qiao F, Liu Q, Wei Z, Qi H, Cui S, An L (2017) Microplastics releasing from personal care and cosmetic products in China. Marine Pollution Bull 123(1–2):122–126
36. Löfroth M, Ghasemimehr M, Falk A, von Steyern PV (2019) Bisphenol A in dental materials–existence, leakage and biological effects. Heliyon 5(5):e01711
37. Martin M, Al S, Derouen T (2014) Possible association between dental sealants and urinary Bisphenol A levels in children warrants additional biomonitoring and safety research. J Evidence-Based Dental Pract 14(4):200–202
38. Murphy F, Ewins C, Carbonnier F, Quinn B (2016) Wastewater treatment works (WwTW) as a source of microplastics in the aquatic environment. Environ Sci Technol 50(11):5800–5808
39. Mulligan S, Kakonyi G, Moharamzadeh K, Thornton SF, Martin N (2018) The environmental impact of dental amalgam and resin-based composite materials. Br Dent J 224(7):542
40. Nasser M (2012) Evidence summary: can plastics used in dentistry act as an environmental pollutant? Can we avoid the use of plastics in dental practice? Br Dent J 212(2):89–91

41. Olea-Serrano N, Fernández M, Pulgar R, Olea-Serrano F (2002) Endocrine disrupting chemicals: harmful substances and how to test them. Cadernos de saúde pública 18:489–494
42. Praveena SM, Shaifuddin SNM, Akizuki S (2018) Exploration of microplastics from personal care and cosmetic products and its estimated emissions to marine environment: an evidence from Malaysia. Mar Pollut Bull 136:135–140
43. Renkema A-M, Al-Assad S, Bronkhorst E, Weindel S, Katsaros C, Lisson JA (2008) Effectiveness of lingual retainers bonded to the canines in preventing mandibular incisor relapse. Am J Orthodontics and Dentofacial Orthopedics: Official Publication of the American Association of Orthodontists, Its Constituent Societies, and the American Board of Orthodontics 134(2):179e1–179e8
44. Ryan PG (2015) A brief history of marine litter research. In: Bergmann M, Gutow L, Klages M (eds) Marine anthropogenic litter, pp 1–25
45. Sakr OM, Alhablain EA (2018) Assessment of the knowledge and attitude of dental students and intern practitioners about the nanotechnology in dentistry at KSA. IAIM 5(10):87–94
46. Saravi ME, Vojdani M, Bahrani F (2012) Evaluation of cellular toxicity of three denture base acrylic resins. J Dent 9(4):180–188
47. Sasaki N, Okuda K, Kato T (2005) Salivary bisphenol—a levels detected by ELISA after restoration with composite resin. J Mater Sci---Mater Med 6:297–300
48. Schweikl H, Spagnuolo G, Schmalz G (2006) Genetic and cellular toxicology of dental resin monomers. J Dent Res 85(10):870–877. https://doi.org/10.1177/154405910608501001
49. Sunitha C, Kailasam V, Padmanabhan S, Chitharanjan AB (2011) Bisphenol A release from an orthodontic adhesive and its correlation with the degree of conversion on varying light-curing tip distances. Am J Orthod Dentofac Orthop 140(2):239–244
50. Tarumi H, Imazato S, Narimatsu M, Matsuo M, Ebisu S (2000) Estrogenicity of fissure sealants and adhesive resins determined by reporter gene assay. J Dent Res 79(11):1838–1843
51. Ter Halle A, Ladirat L, Martignac M, Mingotaud AF, Boyron O, Perez E (2017) To what extent are microplastics from the open ocean weathered? Environ Pollut 227:167–174
52. Van Landuyt KL, Nawrot T, Geebelen B, De Munck J, Snauwaert J, Yoshihara K, Scheers H, Godderis L, Hoet P, Van Meerbeek B (2011) How much do resin-based dental materials release? A meta-analytical approach. Dental Mater: Off Publ Acad Dental Mater 27(8):723–747
53. Vom Saal FS, Hughes C (2005) An extensive new literature concerning low-dose effects of bisphenol A shows the need for a new risk assessment. Environ Health Perspect 113(8):926–933
54. Urban VM, Machado AL, Oliveira RV, Vergani CE, Pavarina AC, Cass QB (2007) Residual monomer of reline acrylic resins: effect of water-bath and microwave post-polymerization treatments. Dent Mater 23(3):363–368
55. Ustabasi GS, Baysal A (2019) Occurrence and risk assessment of microplastics from various toothpastes. Environ Monit Assess 191(7):438
56. Ustabasi GS, Baysal A (2020) Bacterial interactions of microplastics extracted from toothpaste under controlled conditions and the influence of seawater. Sci Total Environ 703:135024
57. Wagner M, Scherer C, Alvarez-muñoz D, Brennholt N, Bourrain X, Buchinger S, Reifferscheid G (2014) Microplastics in freshwater ecosystems: what we know and what we need to know. Environ Sci Europe 26(1):12

Printed in the United States
by Baker & Taylor Publisher Services